AF713027

ALL ABOUT BUTTERFLIES

SCIENCE, SOCIAL STUDIES, ECOLOGY,
GEOGRAPHY, LANGUAGE ARTS,
LOGIC & DRAWING

200 FUN-SCHOOLING ACTIVITIES

HOMESCHOOLING HANDBOOK

By: Notika Pashchenko & Sarah Janisse Brown

Name:

ABOUT THIS BOOK:

200 Activities for students who are interested in butterflies, flowers, ecology, science and nature.

Children ages 9+ should be able to work through this book with minimal assistance; younger children will need help with reading, writing and research. This workbook is fun for all ages, and can be used in collaboration with multiple siblings as a family project. Help your child choose safe research materials and documentaries. Use 2 to 8 pages per day as part of your homeschooling plan.

This workbook covers almost every required subject; just add math and history!

This research handbook is packed with learning prompts and fun activities that all revolve around butterflies. Homeschooling is so much fun when students are studying their passions. If your child is curious about butterflies, this is an excellent tool for their research and exploration.

THIS CURRICULUM INCLUDES:

- Instructions
- Language Arts
- Science
- Nature Study
- Geography
- Art & Drawing
- Documentaries
- Research & Library Skills
- Crafts
- Cursive Writing
- Copywork
- Reading & Spelling
- Math Time
- Creative Writing
- Recycling Skills
- Butterfly Types
- Movie Time
- Coloring
- Mazes
- About Fun-Schooling

INSTRUCTIONS FOR THIS FUN-SCHOOLING RESEARCH HANDBOOK

ABOUT THIS RESEARCH HANDBOOK:

This Fun-Schooling Research Handbook gives you, your child, or student a simple framework to make research engaging, manageable, and creative.

Instead of traditional worksheets, this book uses guided prompts to help learners explore topics they are curious about while practicing real research skills across multiple subjects.

GETTING STARTED:

1. Choose a Topic. Select a subject or question your learner wants to explore.
2. Gather information. Use a variety of resources, such as: library books, parent approved websites, videos, documentaries, or movies.
3. Respond to prompts. Each page includes research-based prompts to guide learners to observe and collect information; think critically about what they learn; and reflect, write, draw or organize ideas in their own way.

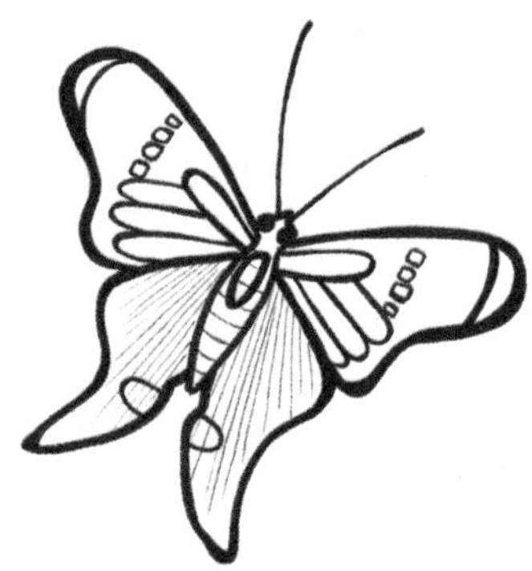

SUPPLIES YOU MAY NEED:

- Library books or home resources
- Documentaries or movies
- Websites
- Pencil or pens
- Colored pencils or markers

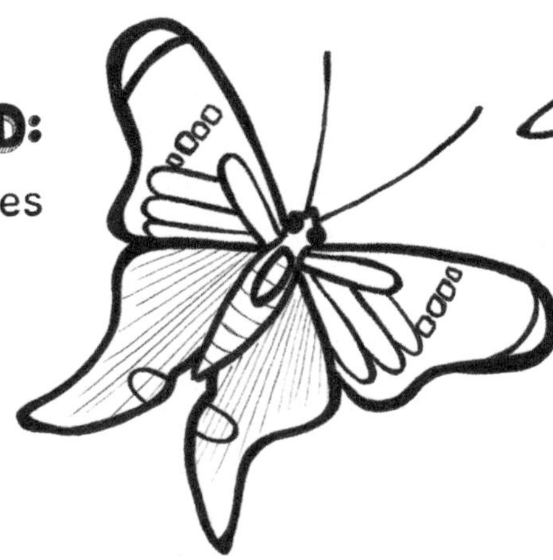

HOW TO SUPPORT YOUR STUDENT:

This book is not about perfect answers or polished pages. It's about sparking curiosity, exploring, and building confidence.

Each page invites students to slow down, observe, and respond in their own way. Responses may include writing, sketching, mapping ideas, or adding color and symbols.

Look for:

- Care and intention in responses
- Curiosity and imagination
- Growing confidence and independence
- Patience and focus while working

If a page feels rushed or incomplete, gently encourage your student to revisit it later or approach the next page or section with more intention. Learning deepens when students feel unhurried and supported.

WHY RESEARCH HANDBOOKS MATTER:

Research is the key that unlocks learning for life.

When students understand how to research, they are no longer dependent on being taught every answer. They gain confidence to:

- Explore new interests independently
- Solve real-world problems
- Learn continuously as the world changes

This handbook is designed to gently build that skill through repeated guided prompts helping students become capable, confident learners.

This book doesn't teach students what to think. It teaches them how to learn.

INSTRUCTIONS

CHOOSE YOUR TOPICS!

Let's learn about butterflies, flowers, gardens and wildlife.

Draw FIVE things you are curious about:

ACTION STEPS:

1. Go to the library or bookstore.
2. Bring home a stack of at least FIVE interesting books about these topics.
3. Choose some that have diagrams, instructions and illustrations.

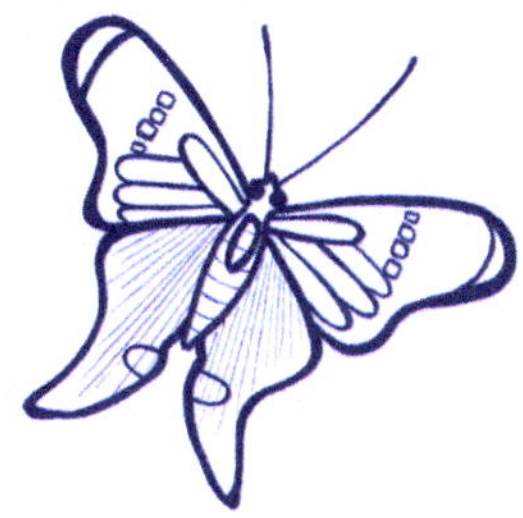

SCHOOL SUPPLIES NEEDED:

Pencils, Colored Pencils & Gel Pens.

PARENTS PLEASE CREATE A LIST OF APPROVED WEBSITES AND RESOURCES FOR YOUR CHILD:

____

____

____

____

____

____

____

____

____

____

____

____

____

____

GO TO THE LIBRARY AND CHOOSE FIVE BOOKS TO USE AS SCHOOL BOOKS!

1. Write down the titles on each book cover below. Extra books can be recorded on pages 218–220.

2. Keep your stack of books in a safe place so you can read a few pages from your books daily.

3. Ask your parent or teacher how many pages to complete each day in this handbook. Five to eight pages is normal for kids your age.

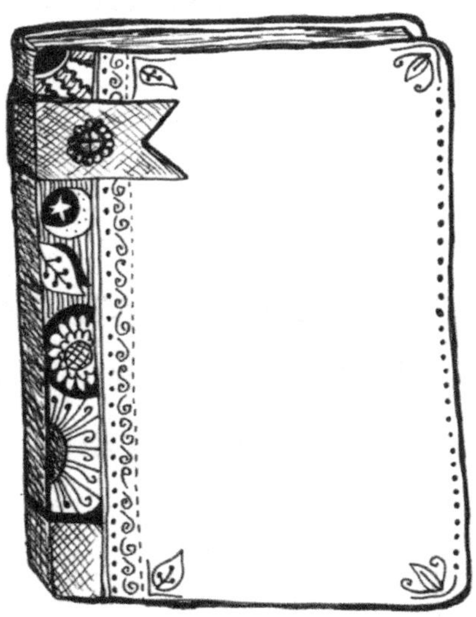

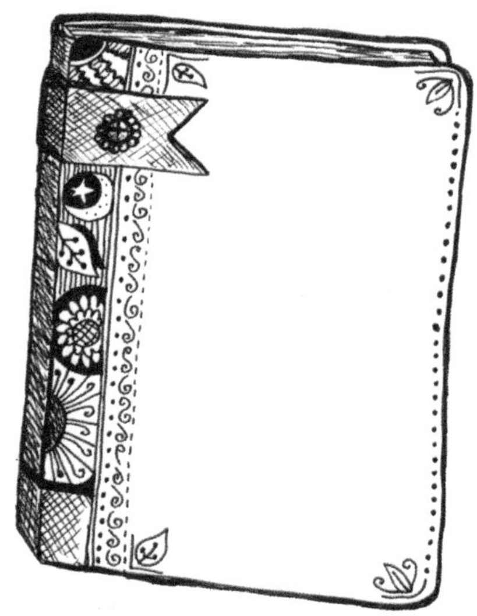

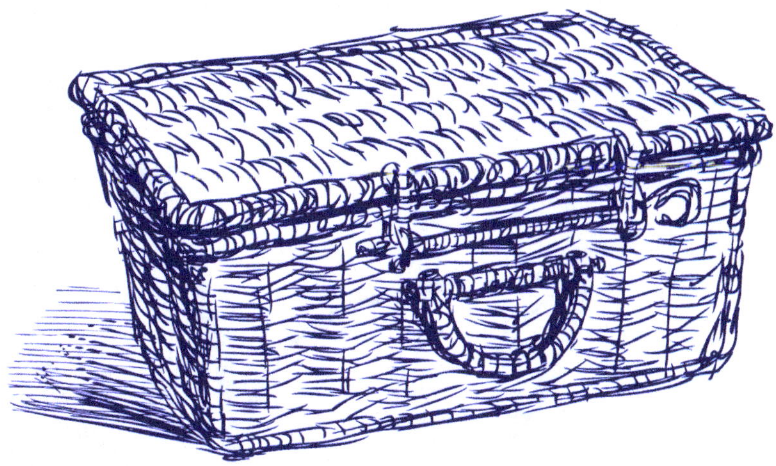

You may choose new books any time.

Flip to the back for more book pages.

Keep all your books in a basket with your pens and pencils.

Have a snack before you start working in this journal.

BUTTERFLY LIFE CYCLE

This butterfly is called a spicebush swallowtail.

1. Egg

2. Caterpillar

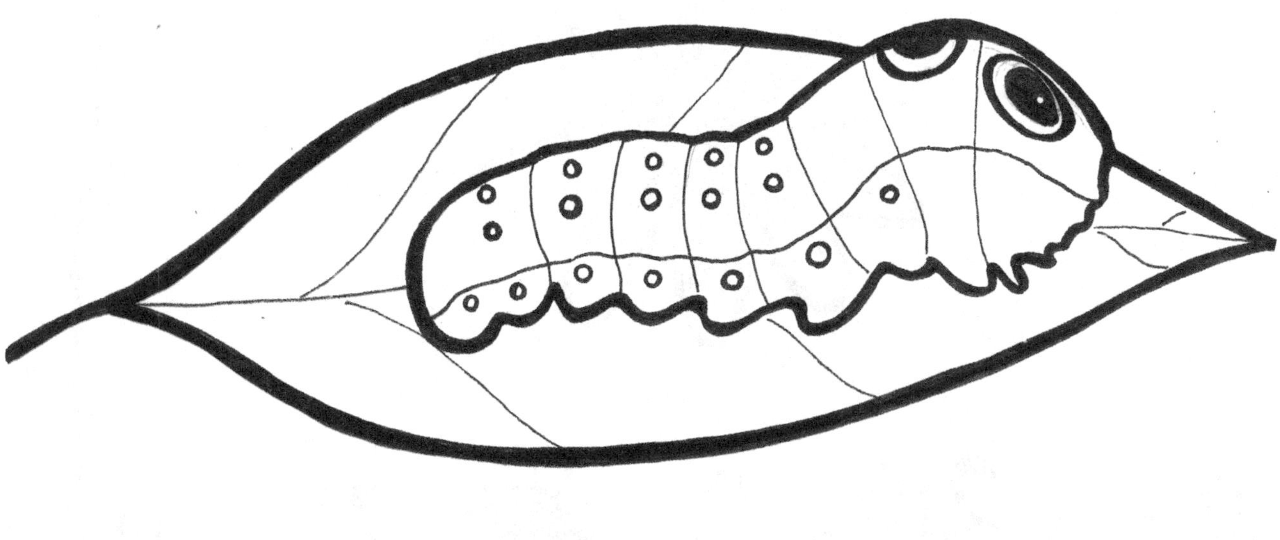

3. Pupa

4. Adult butterfly

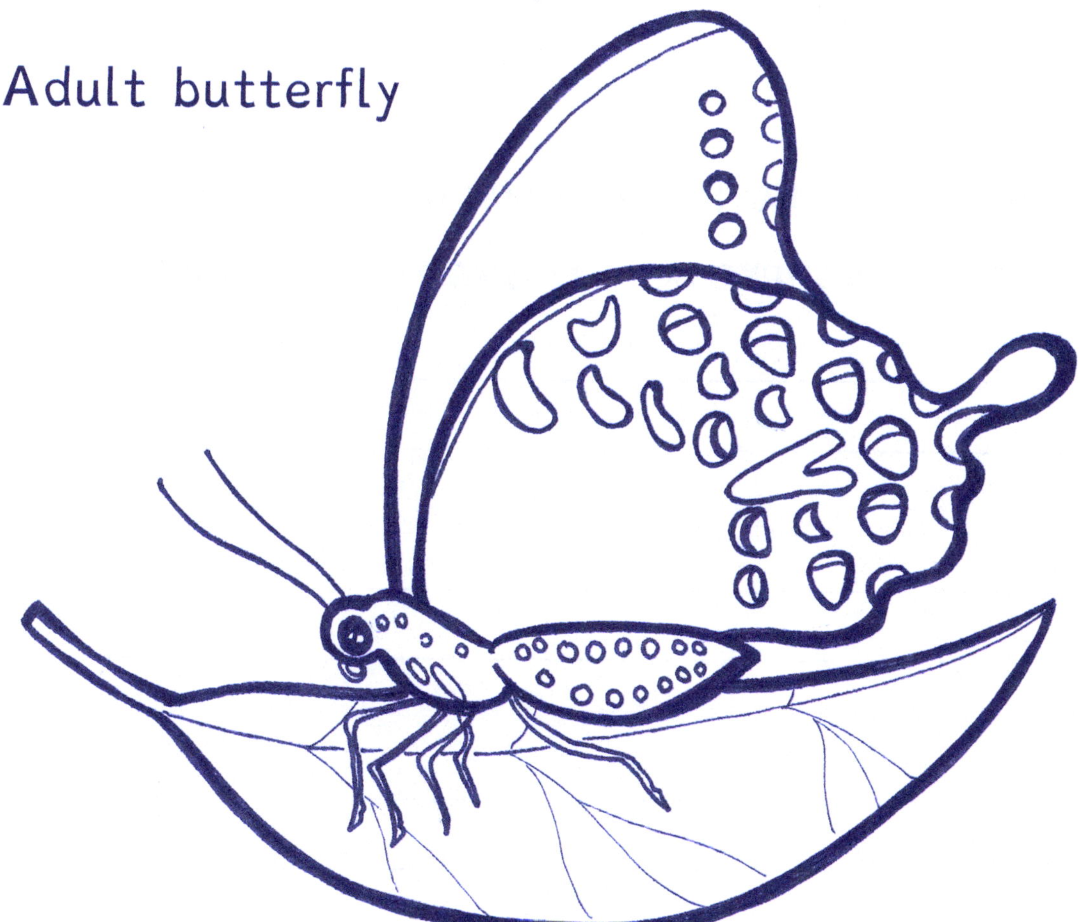

ALL ABOUT
ATLAS MOTH

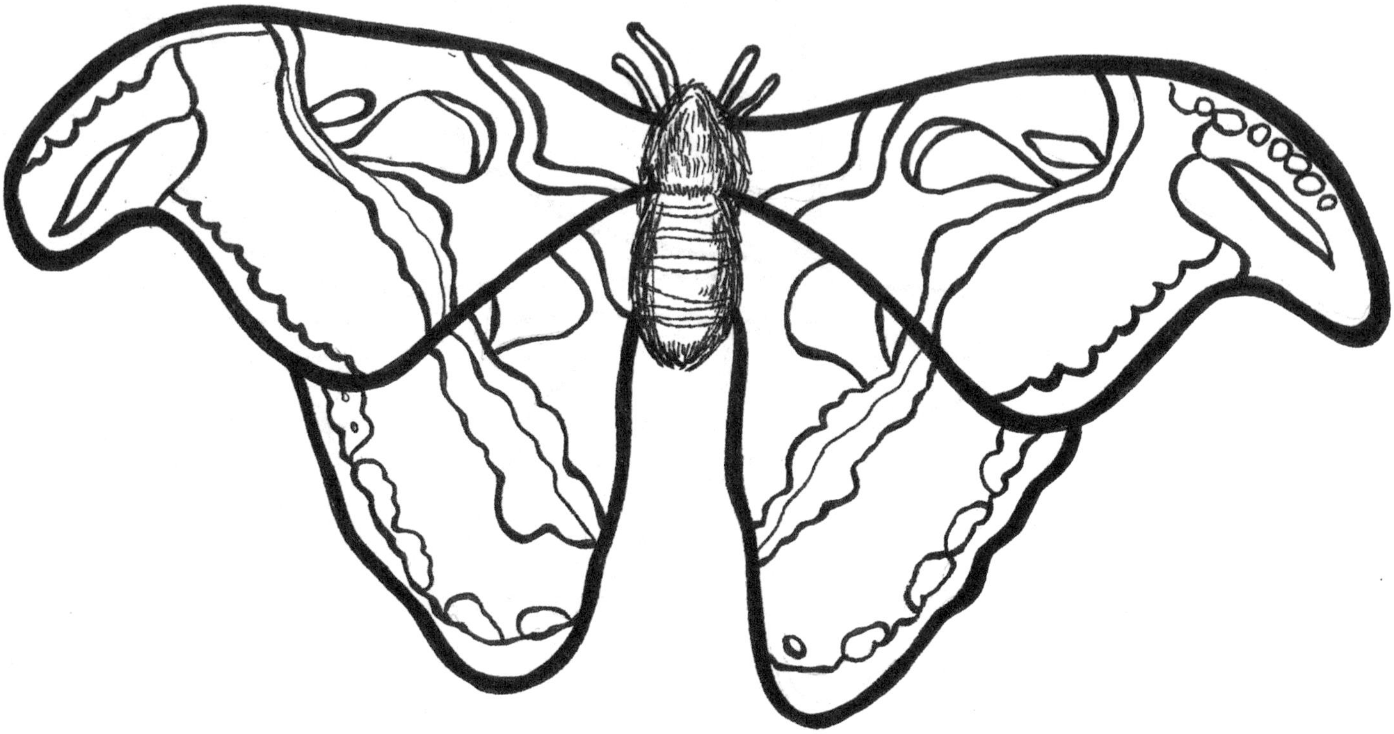

LOOK UP THIS MOTH AND COLOR ACCURATELY

AND WRITE DOWN THREE INTERESTING FACTS ABOUT IT.

1. ____

____

2. ____

____

3. ____

____

RESEARCH & DISCOVERIES

USE LIBRARY BOOKS, ENCYCLOPEDIAS OR THE INTERNET TO LEARN MORE.

Color the parts of the world where this animal lives.

DRAW ONE OF MY HOST PLANTS	DRAW MY FOOD	DRAW MY ENEMIES

CREATIVE WRITING

Write a story or poem about the illustration on the right.

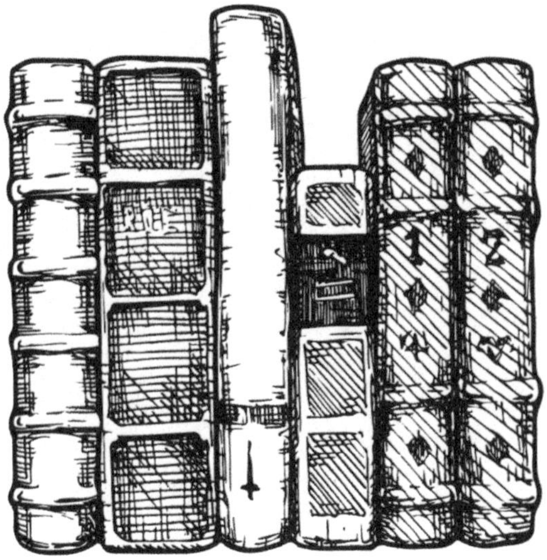

Today I will read for

15 30 45 60

MINUTES

READING TIME

Ask your teacher to help you decide how many books to read from each day. #______

Write and draw about what you are reading.

BACKYARD SCIENCE

NATURE WALK & NATURE STUDY

Draw or write about the things you see outside today.

If possible, find a butterfly or moth to draw!

COMPLETE & COLOR

A B C D E F G

H I J K L M

N O P Q R S T

U V W X Y Z

a b c d e f g h

i j k l m n o p

q r s t u v w x

y z 1 2 3 4 5 6 7 8 9 0

CURSIVE WRITING PRACTICE

MATH TIME

Use this page to practice math!

MOVIE TIME

Watch a video or documentary about BUTTERFLIES.

TITLE:______________________

____

RATING:

Draw Your Favorite Scenes:

ALL ABOUT MONARCH BUTTERFLY

LOOK UP THIS BUTTERFLY AND COLOR ACCURATELY AND WRITE DOWN THREE INTERESTING FACTS ABOUT IT.

1. ______________________________

____

2. ______________________________

____

3. ______________________________

____

RESEARCH & DISCOVERIES

USE LIBRARY BOOKS, ENCYCLOPEDIAS OR THE INTERNET TO LEARN MORE.

Color the parts of the world where this animal lives.

DRAW ONE OF MY HOST PLANTS	DRAW MY FOOD	DRAW MY ENEMIES

CREATIVE WRITING

Write a story or poem about the illustration on the right.

Queen Alexandra's Birdwing Butterfly

Often, male and female butterflies of the same kind have different color, shape and design. Color the butterflies and study how difference in their wings.

Queen Alexandra's Birdwing Butterfly male

Queen Alexandra's Birdwing Butterfly female

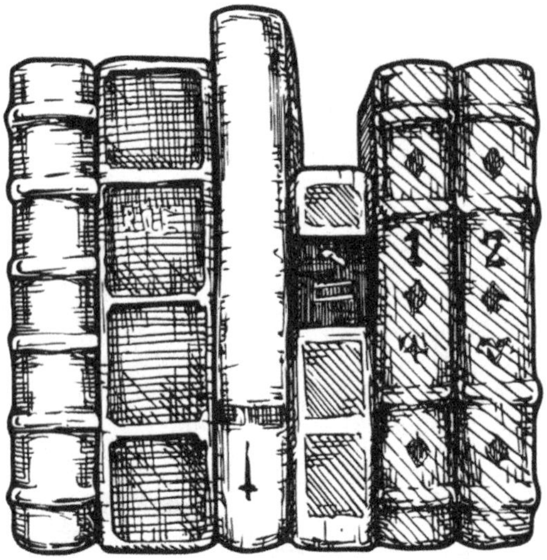

Today I will read for

15 30 45 60

MINUTES

READING

TIME

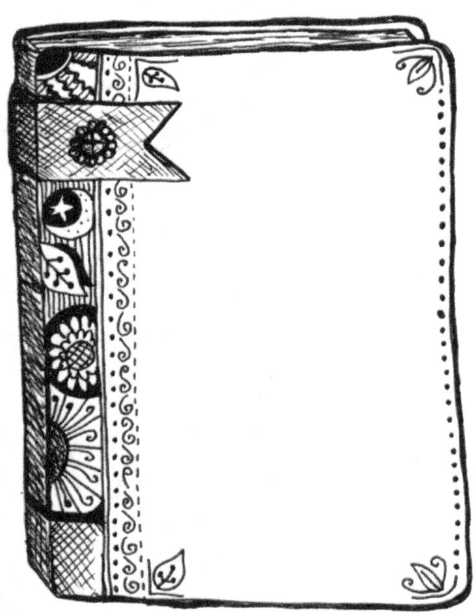

Write and draw about what you are reading.

BACKYARD SCIENCE
NATURE WALK & NATURE STUDY

Draw or write about the things you see outside today.

If possible, find a butterfly or moth to draw!

SPELLING TIME

Pick a Letter ________

Look in your nature and butterfly books for words that start with this letter.

Write ten spelling words.

1. ________________________________

2. ________________________________

3. ________________________________

4. ________________________________

5. ________________________________

6. ________________________________

7. ________________________________

8. ________________________________

9. ________________________________

10. ________________________________

LISTENING TIME

IDEAS: MUSIC, PODCASTS, & AUDIO BOOKS

I am listening to: ______________________________

ALL ABOUT WESTERN PYGMY BLUE BUTTERFLY

LOOK UP THIS BUTTERFLY AND COLOR ACCURATELY AND WRITE DOWN THREE INTERESTING FACTS ABOUT IT.

1. ______________________________

____

2. ______________________________

____

3. ______________________________

____

RESEARCH & DISCOVERIES

USE LIBRARY BOOKS, ENCYCLOPEDIAS OR THE INTERNET TO LEARN MORE.

Color the parts of the world where this animal lives.

DRAW ONE OF MY HOST PLANTS	DRAW MY FOOD	DRAW MY ENEMIES

CREATIVE WRITING

Write a story or poem about the illustration on the right.

Color the butterfly and add to the background.

COLOR

Use a smooth black marker to complete the drawing

DRAW THE MISSING PARTS

MATH TIME

Use this page to practice math!

MOVIE TIME

Watch a video, cartoon or documentary about RECYCLING.

TITLE:______________________

____

Draw Your Favorite Scenes:

RATING:

REDUCE, REUSE, RECYCLE

Draw 3 things that can be put in this bin.

Draw 3 things that can be made of recycled paper

REDUCE, REUSE, RECYCLE

Draw a poster about saving trees.

CLIP ART

Use stickers, pictures from old magazines, or print out pictures of butterflies and flowers from the internet and paste them on these pages.

COPYWORK

Copy a verse, quote or line of poetry:

TITLE: ____________________

Today I will read for

15 30 45 60

MINUTES

READING TIME

Write and draw about what you are reading.

BACKYARD SCIENCE
NATURE WALK & NATURE STUDY

Draw or write about the things you see outside today.

If possible, find a butterfly or moth to draw!

DRAWING TIME

Practice drawing butterflies, bugs, flowers and plants.

Find a way through the maze

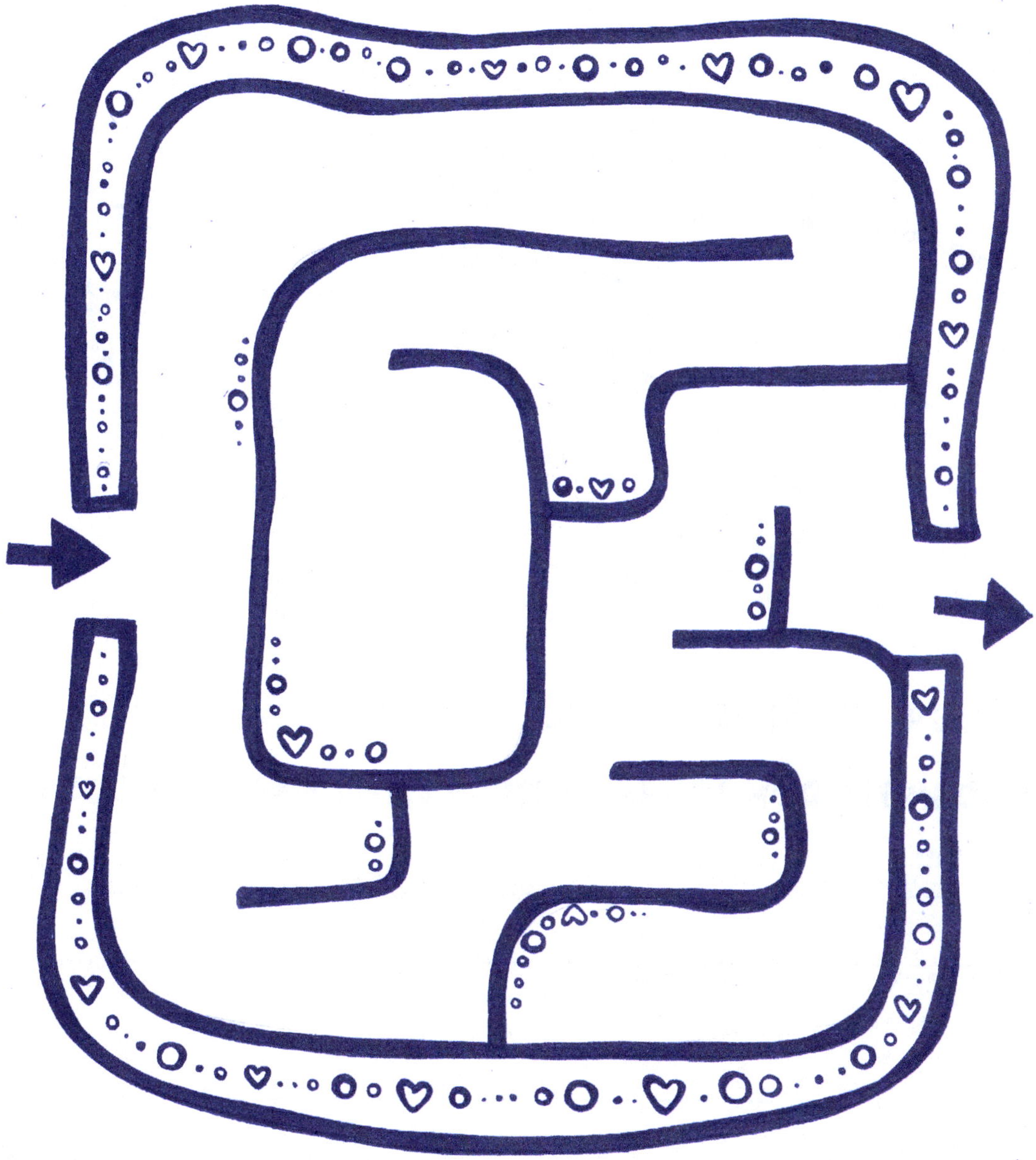

ALL ABOUT
ZEBRA LONGWING BUTTERFLY

LOOK UP THIS BUTTERFLY AND COLOR ACCURATELY AND WRITE DOWN THREE INTERESTING FACTS ABOUT IT.

1. ____
____

2. ____
____

3. ____
____

RESEARCH & DISCOVERIES

USE LIBRARY BOOKS, ENCYCLOPEDIAS OR THE INTERNET TO LEARN MORE.

Color the parts of the world where this animal lives.

DRAW ONE OF MY HOST PLANTS	DRAW MY FOOD	DRAW MY ENEMIES

CREATIVE WRITING

Write a story or poem about the illustration on the right.

MATH TIME

Use this page to practice math!

MOVIE TIME

Watch a video or documentary about BEES.

TITLE:______________________

____

RATING:

Draw Your Favorite Scenes:

Color one half of a butterfly with poster paint or craft paint.

While the paint is still wet, close the book and gently press pages together to transfer color to the other wing.

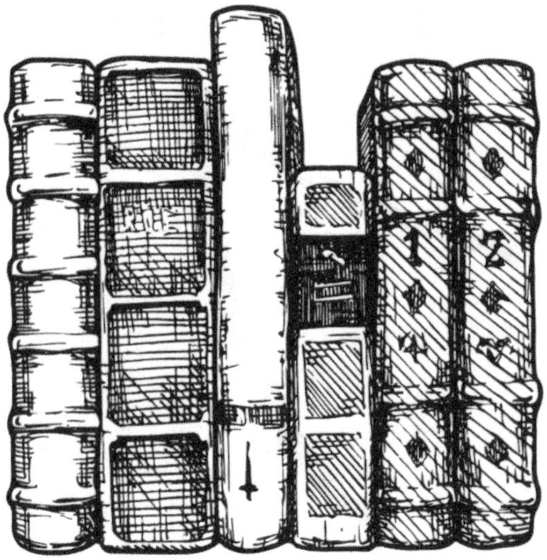

Today I will read for

15 30 45 60

MINUTES

READING

TIME

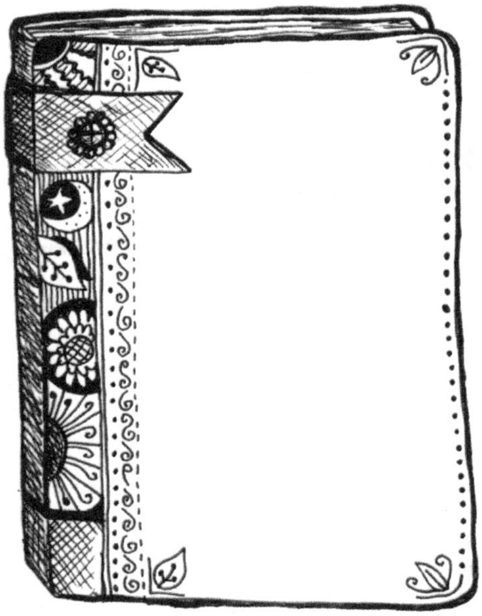

Write and draw about what you are reading.

BACKYARD SCIENCE

NATURE WALK & NATURE STUDY

Draw or write about the things you see outside today.

If possible, find a butterfly or moth to draw!

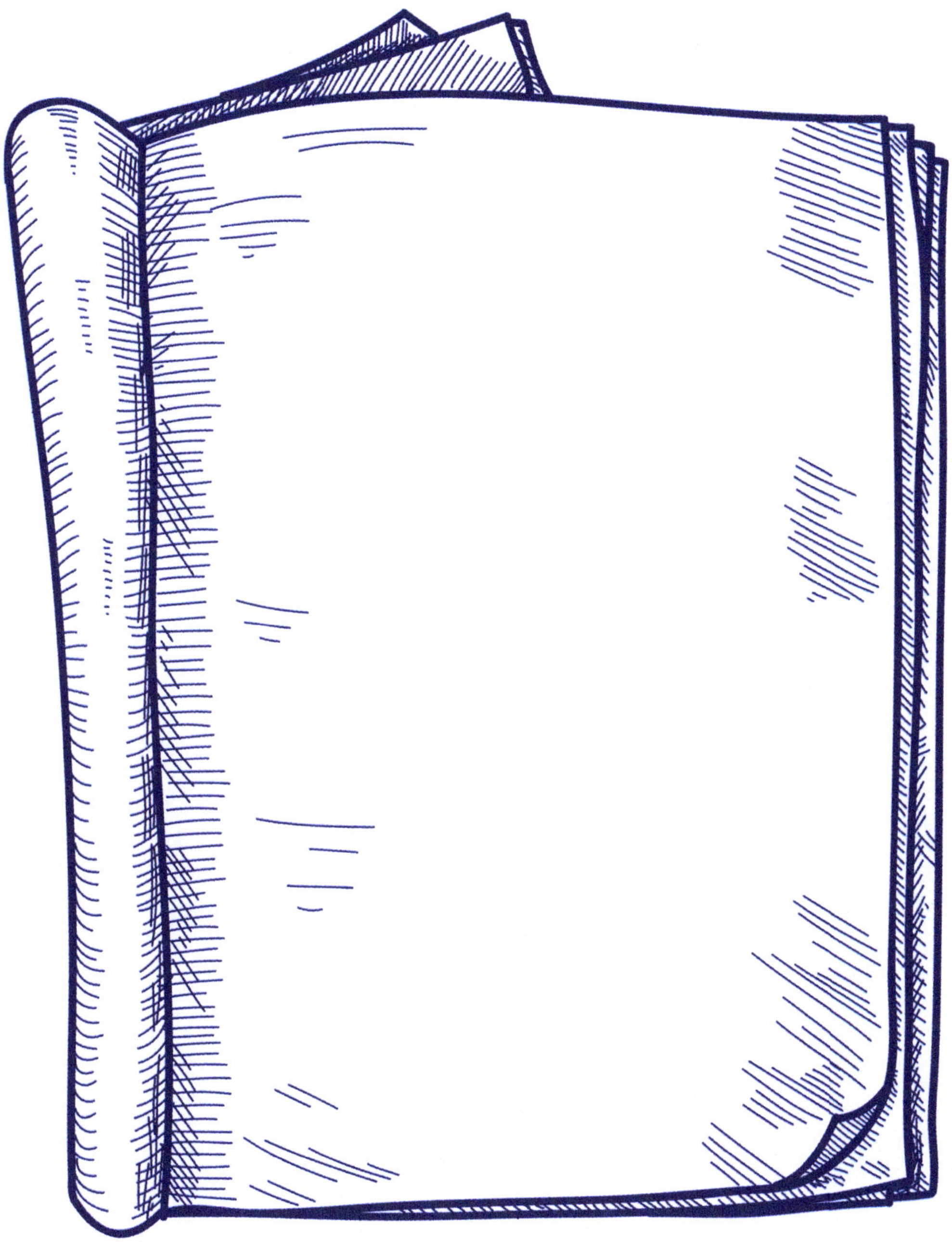

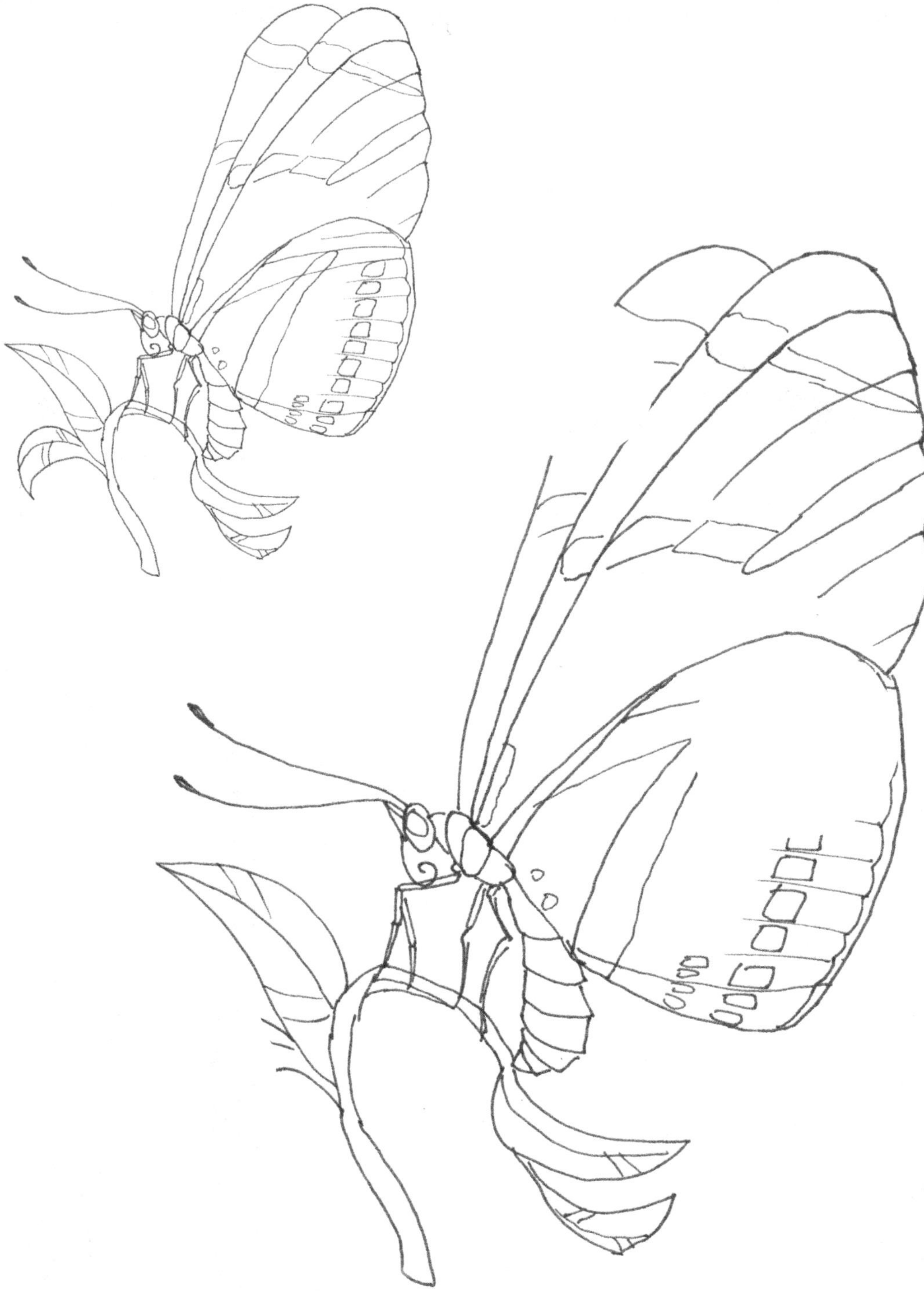

COMPLETE & COLOR

A B C D E F G

H I J K L M

N O P Q R S T

U V W X Y Z

a b c d e f g h

i j k l m n o p

q r s t u v w x

y z 1 2 3 4 5 6 7 8 9 0

CURSIVE WRITING PRACTICE

ALL ABOUT
88 BUTTERFLY

LOOK UP THIS BUTTERFLY AND COLOR ACCURATELY AND WRITE DOWN THREE INTERESTING FACTS ABOUT IT.

1.____

____

2.____

____

3.____

____

RESEARCH & DISCOVERIES

USE LIBRARY BOOKS, ENCYCLOPEDIAS OR THE INTERNET TO LEARN MORE.

Color the parts of the world where this animal lives.

DRAW ONE OF MY HOST PLANTS	DRAW MY FOOD	DRAW MY ENEMIES

CREATIVE WRITING

Write a story or poem about the illustration on the right.

Color the butterfly and add to the background.

Use a smooth black marker to complete the drawing

DRAW THE MISSING PARTS

MATH TIME

Use this page to practice math!

MOVIE TIME

Watch a video, cartoon or documentary about OCEAN POLLUTION .

TITLE:________________________

____

Draw Your Favorite Scenes:

RATING:

SPELLING TIME

Pick a Letter ________

Look in your nature and butterfly books for words that start with this letter.

Write ten spelling words.

1.______________________________

2.______________________________

3.______________________________

4.______________________________

5.______________________________

6.______________________________

7.______________________________

8.______________________________

9.______________________________

10.______________________________

LISTENING TIME

IDEAS: MUSIC, PODCASTS, & AUDIO BOOKS

I am listening to: ______________________________

HOST PLANTS

Host plants are plants that support life cycle of different insects. The host plant provides a place for butterflies or insects to lay eggs. These plants provide food, nectar and shelter for caterpillars, butterflies and other insects.

This plant is milkweed.

It is a host plant of the monarch butterfly.

It is important to preserve forests and meadows because butterflies depend on them for their life. Read and find out which butterflies are living close to your home. Find out about their host plants. You can attract butterflies if you plant some of these plants in your garden.

Draw one of the host plants that grows in your neighborhood.

ALL ABOUT
LONG-TAILED SKIPPER BUTTERFLY

LOOK UP THIS BUTTERFLY AND COLOR ACCURATELY AND WRITE DOWN THREE INTERESTING FACTS ABOUT IT.

1. ____

____

2. ____

____

3. ____

____

RESEARCH & DISCOVERIES

USE LIBRARY BOOKS, ENCYCLOPEDIAS OR THE INTERNET TO LEARN MORE.

Color the parts of the world where this animal lives.

DRAW ONE OF MY HOST PLANTS	DRAW MY FOOD	DRAW MY ENEMIES

CREATIVE WRITING

Write a story or poem about the illustration on the right.

CLIP ART

Use stickers, pictures from old magazines, or print out pictures of butterflies and flowers from the internet and paste them on these pages.

COPYWORK

Copy a verse, quote or line of poetry:

TITLE: ____________________

DRAWING TIME

Practice drawing butterflies, bugs, flowers and plants.

Find a way through the maze

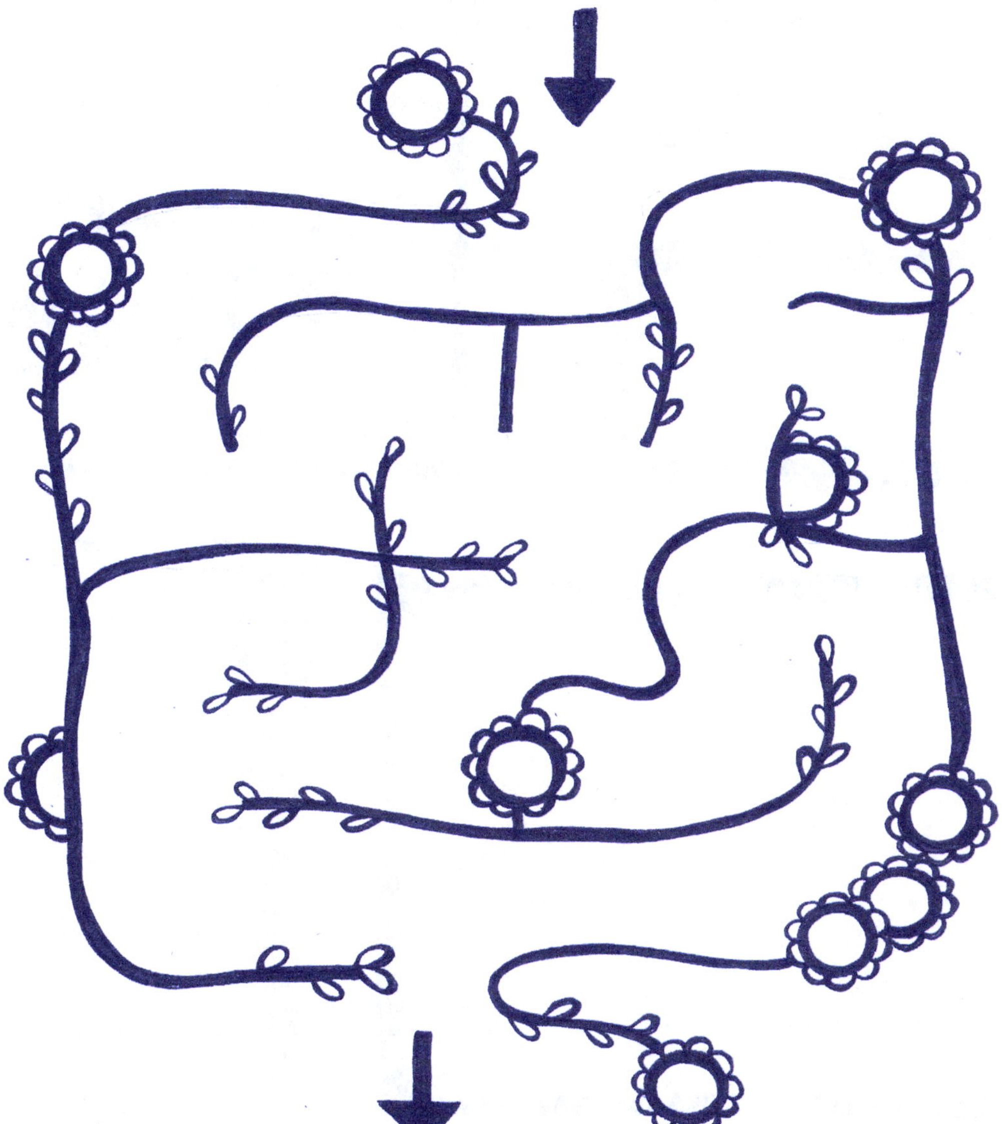

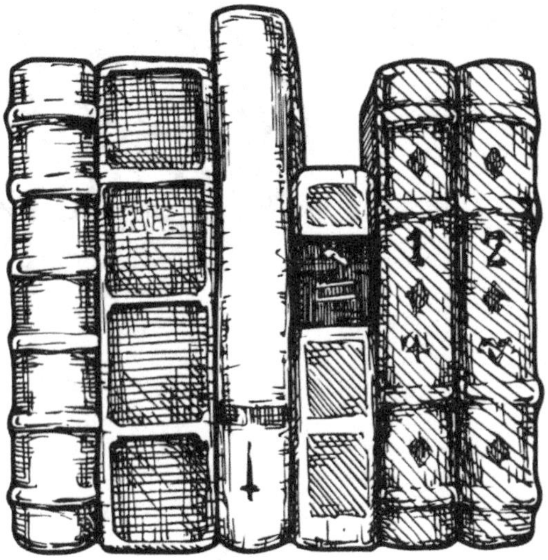

Today I will read for

15 30 45 60

MINUTES

READING TIME

Write and draw about what you are reading.

BACKYARD SCIENCE
NATURE WALK & NATURE STUDY

Draw or write about the things you see outside today.

If possible, find a butterfly or moth to draw!

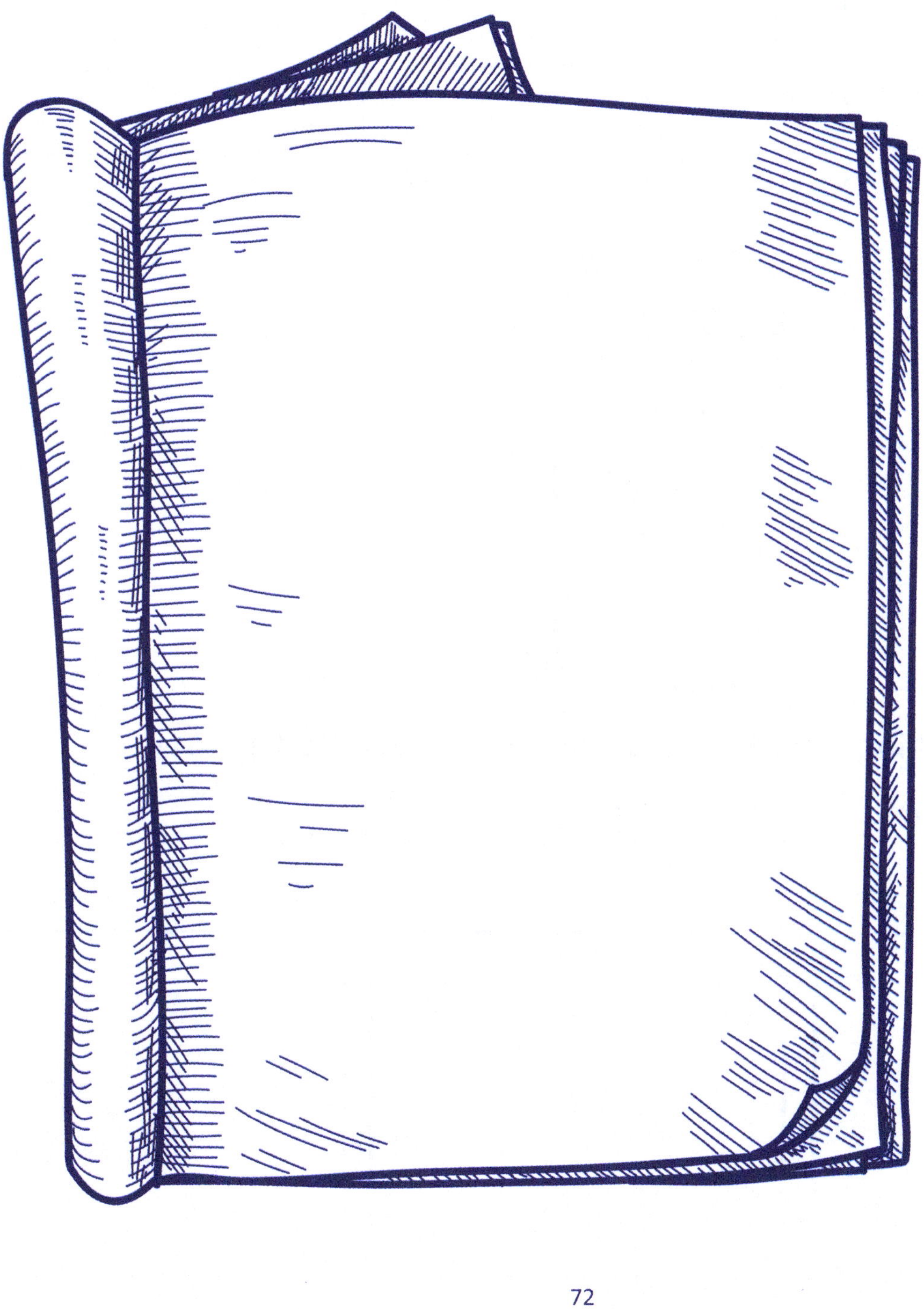

REDUCE, REUSE, RECYCLE

Draw **3** things that can be put in this bin

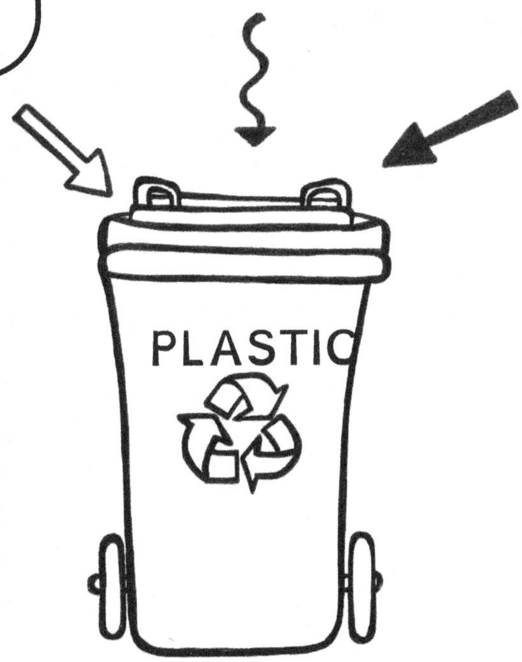

Draw **3** things that can be made of recycled plastic

REDUCE, REUSE, RECYCLE

Draw a picture to illustrate why it is important to use less plastic.

ALL ABOUT COMMON BLUE BUTTERFLY

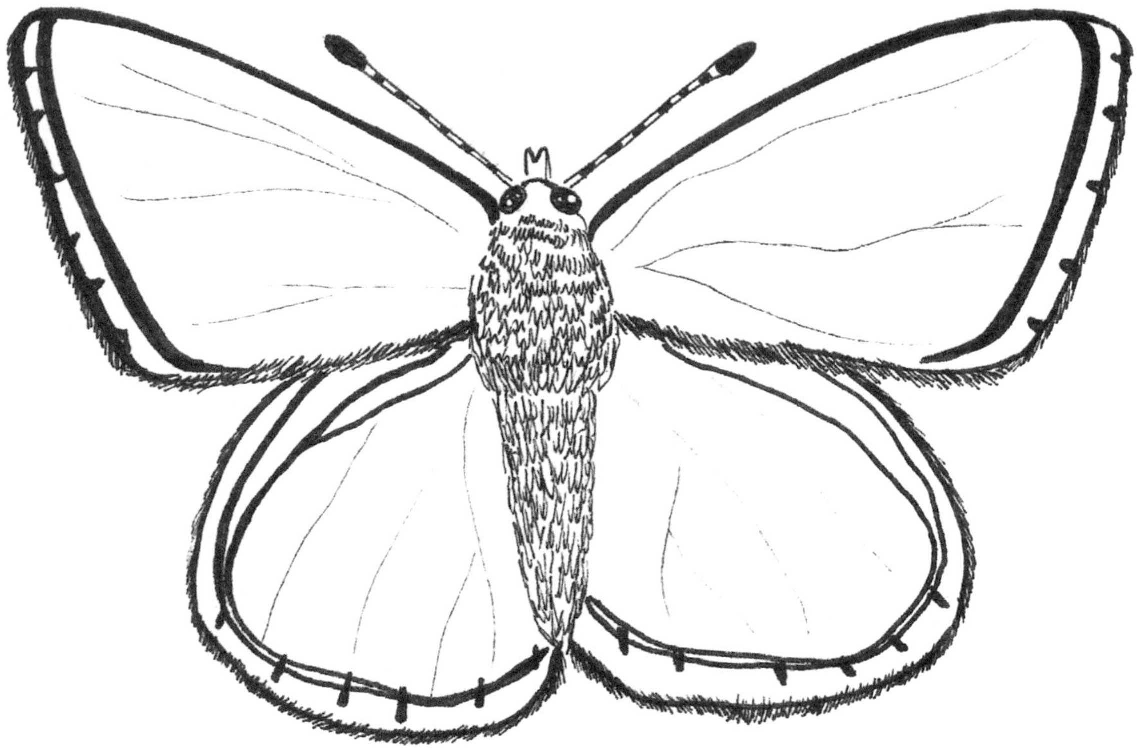

LOOK UP THIS BUTTERFLY AND COLOR ACCURATELY AND WRITE DOWN THREE INTERESTING FACTS ABOUT IT.

1. ____

____

2. ____

____

3. ____

____

RESEARCH & DISCOVERIES

USE LIBRARY BOOKS, ENCYCLOPEDIAS OR THE INTERNET TO LEARN MORE.

Color the parts of the world where this animal lives.

DRAW ONE OF MY HOST PLANTS	DRAW MY FOOD	DRAW MY ENEMIES

CREATIVE WRITING

Write a story or poem about the illustration on the right.

MATH TIME

Use this page to practice math!

MOVIE TIME

Watch a video or documentary about FLOWERS OR PLANTS.

TITLE:________________________

____

RATING:

Draw Your Favorite Scenes:

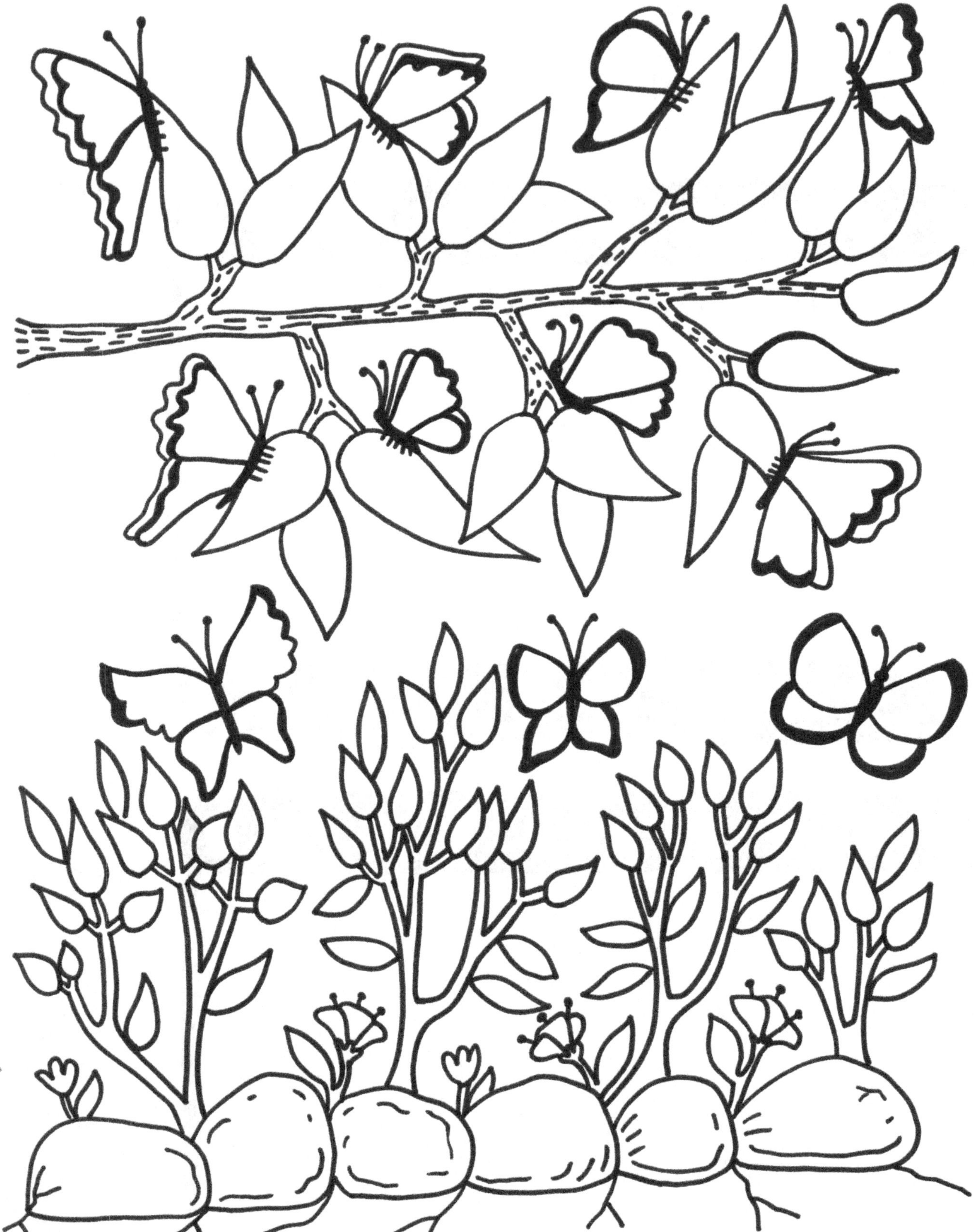

COMPLETE & COLOR

A B C D E F G

H I J K L M

N O P Q R S T

U V W X Y Z

a b c d e f g h

i j k l m n o p

q r s t u v w x

y z 1 2 3 4 5 6 7 8 9 0

CURSIVE WRITING PRACTICE

POLLINATION

Some plants and flowers depend on pollination to grow and produce fruits. Pollen is a fine powder that has to be transferred from one flower to another. Bees, butterflies, other insects, birds and some bats help with pollination. As pollinators visit different flowers, they transfer pollen.

Can you find out where nectar and pollen are in the lily?

Today I will read for

15 30 45 60

MINUTES

READING TIME

Write and draw about what you are reading.

POLLINATION EXPERIMENT

You will need: flour, food coloring or poster paint, cotton swab, magnifying glass.

Conduct an experiment to see how pollination works. Make colorful powder with wheat flour (substitution: potato starch or powdered sugar). You will need 3 spoonfuls of powder to mix with different colors. Mix one spoon of powder with a drop of food coloring or poster paint. Leave it for a few minutes to dry. Place colored powder in the center of each flower. Use a cotton swab or pipe cleaner and dip it in each flower's "pollen" the same way like legs of bees and butterfly would touch the flower's pollen. Use magnifying glass to examine every flower.

SPELLING TIME

Pick a Letter ________

Look in your nature and butterfly books for words that start with this letter.

Write ten spelling words.

1.________________________________

2.________________________________

3.________________________________

4.________________________________

5.________________________________

6.________________________________

7.________________________________

8.________________________________

9.________________________________

10.________________________________

LISTENING TIME

IDEAS: MUSIC, PODCASTS, & AUDIO BOOKS

I am listening to: ______________________________

ALL ABOUT
MOURNING CLOAK

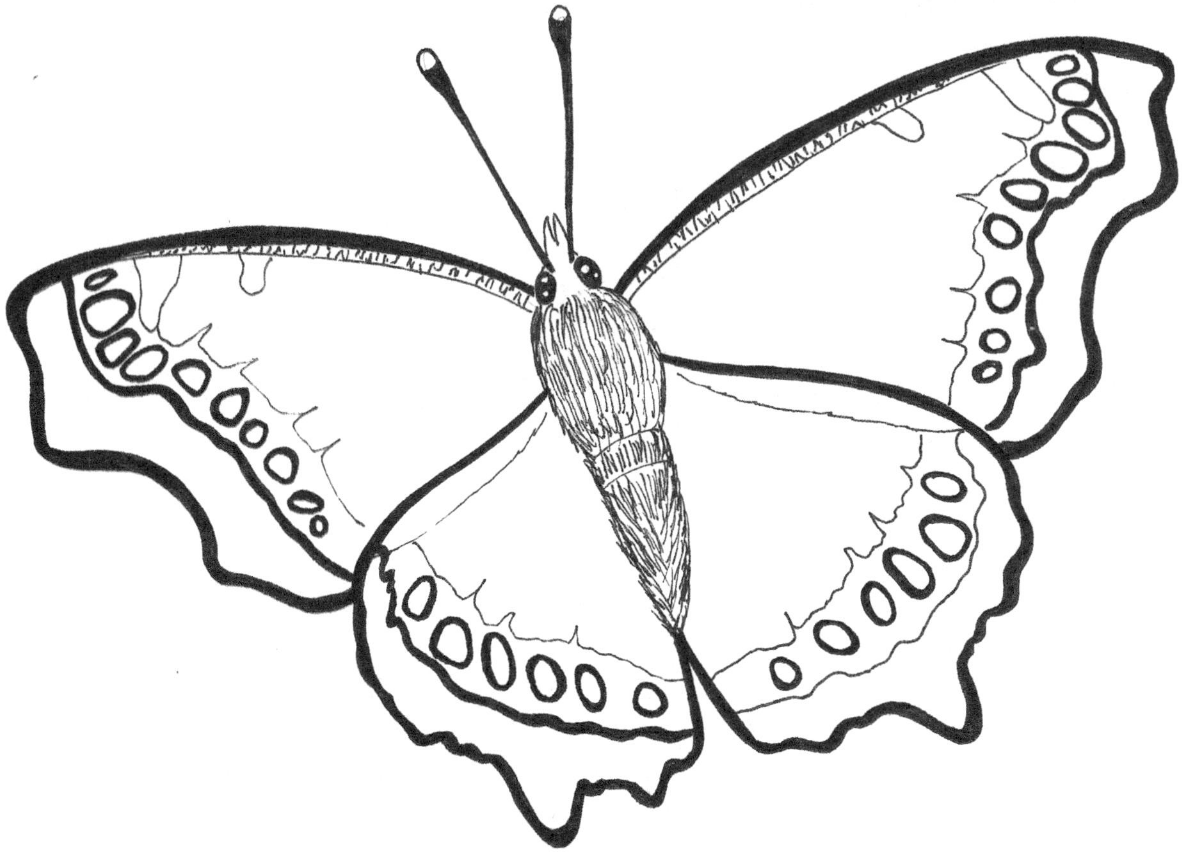

LOOK UP THIS BUTTERFLY AND COLOR ACCURATELY AND WRITE DOWN THREE INTERESTING FACTS ABOUT IT.

1. ____

____

2. ____

____

3. ____

____

RESEARCH & DISCOVERIES

USE LIBRARY BOOKS, ENCYCLOPEDIAS OR THE INTERNET TO LEARN MORE.

Color the parts of the world where this animal lives.

DRAW ONE OF MY HOST PLANTS	DRAW MY FOOD	DRAW MY ENEMIES

CREATIVE WRITING

Write a story or poem about the illustration on the right.

Color the butterfly and add to the background.

Use a smooth black marker to complete the drawing

DRAW THE MISSING PARTS

MATH TIME

Use this page to practice math!

MOVIE TIME

Watch a video, cartoon or documentary about PROTECTING TREES.

TITLE:________________________

____

RATING:

Draw Your Favorite Scenes:

CLIP ART

Use stickers, pictures from old magazines, or print out pictures of butterflies and flowers from the internet and paste them on these pages.

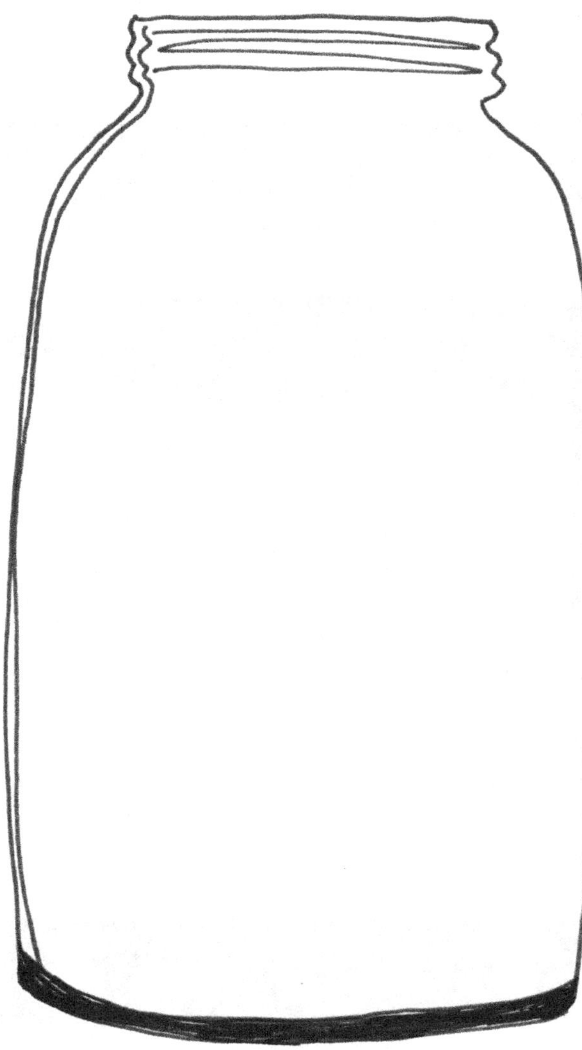

COPYWORK

Copy a verse, quote or line of poetry:

TITLE: ______________________

DRAWING TIME

Practice drawing butterflies, bugs, flowers and plants.

Find a way through the maze

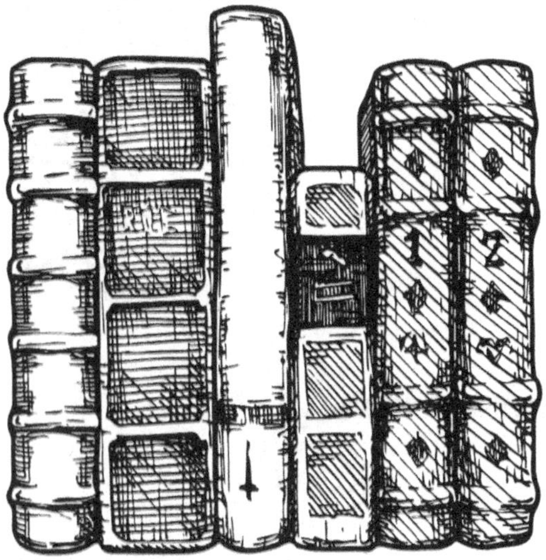

Today I will read for

15 30 45 60

MINUTES

READING TIME

Write and draw about what you are reading.

BACKYARD SCIENCE

NATURE WALK & NATURE STUDY

Draw or write about the things you see outside today.

If possible, find a butterfly or moth to draw!

COMPLETE & COLOR

A B C D E F G
H I J K L M
N O P Q R S T
U V W X Y Z

a b c d e f g h
i j k l m n o p
q r s t u v w x
y z 1 2 3 4 5 6 7 8 9 0

CURSIVE WRITING PRACTICE

ALL ABOUT CEANOTHUS SILK MOTH

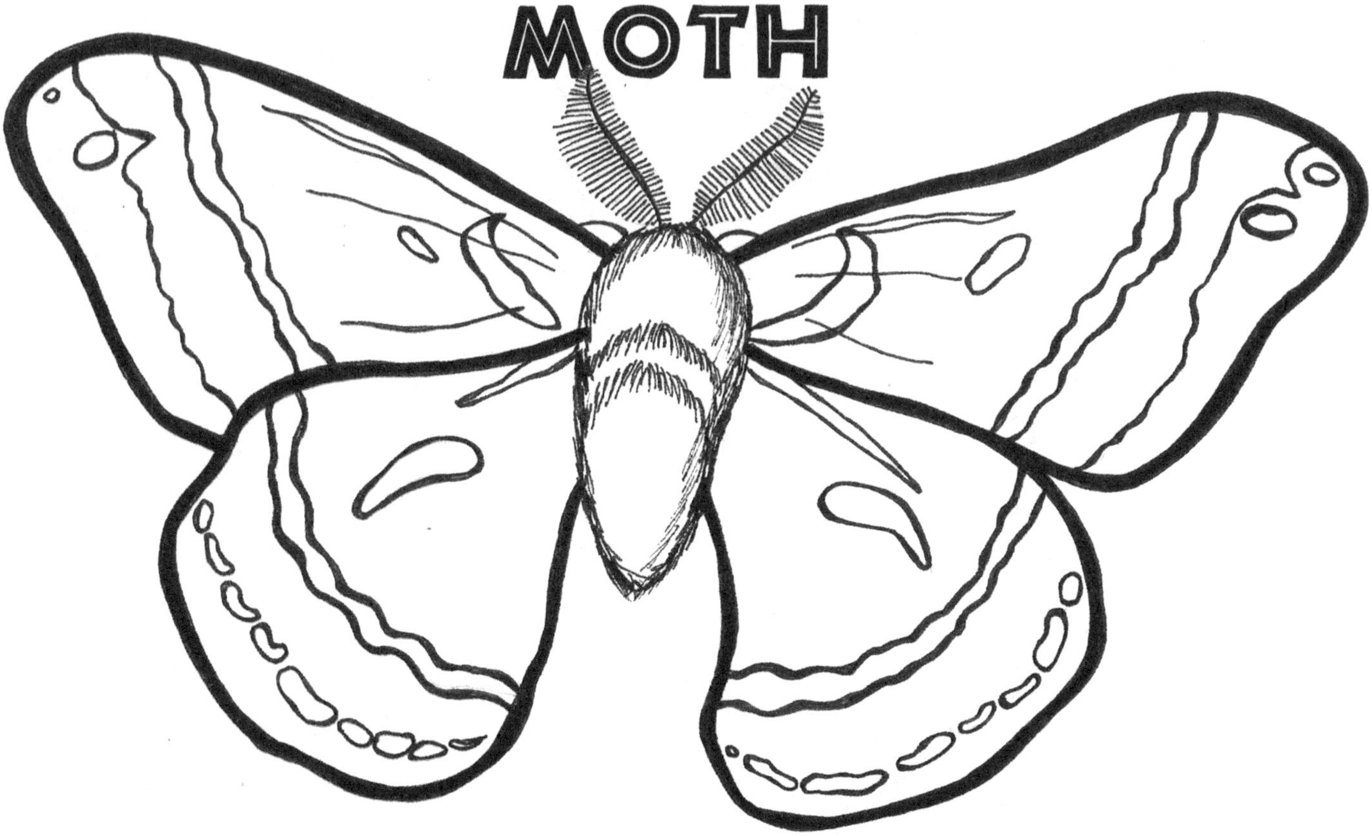

LOOK UP THIS MOTH AND COLOR ACCURATELY

AND WRITE DOWN THREE INTERESTING FACTS ABOUT IT.

1.____

____

2.____

____

3.____

____

RESEARCH & DISCOVERIES

USE LIBRARY BOOKS, ENCYCLOPEDIAS OR THE INTERNET TO LEARN MORE.

Color the parts of the world where this animal lives.

DRAW ONE OF MY HOST PLANTS	DRAW MY FOOD	DRAW MY ENEMIES

CREATIVE WRITING

Write a story or poem about the illustration on the right.

REDUCE, REUSE, RECYCLE

Draw 3 things that can be recycled in this bin

Draw 3 things that can be made of recycled glass

REDUCE, REUSE, RECYCLE

Draw a picture to illustrate how to reduce the amount of waste.

MATH TIME

Use this page to practice math!

MOVIE TIME

Watch a video or documentary about BUTTERFLIES OF AMAZON.

TITLE:________________________

____

RATING:

Draw Your Favorite Scenes:

NATURE CRAFT

For this craft you will need glue and natural materials like grass, leaves and flowers that you can find outside. Cut the butterfly on the next page from the journal. Collect some beautiful leaves and flowers. Apply glue on the butterfly wings and decorate them with the materials you found.

Draw your favorite flower

SPELLING TIME

Pick a Letter ________

Look in your nature and butterfly books for words that start with this letter.

Write ten spelling words.

1. ________________________________
2. ________________________________
3. ________________________________
4. ________________________________
5. ________________________________
6. ________________________________
7. ________________________________
8. ________________________________
9. ________________________________
10. ________________________________

CLIP ART

Use stickers, pictures from old magazines, or print out pictures of butterflies and flowers from the internet and paste them on these pages.

COPYWORK

Copy a verse, quote or line of poetry:

TITLE: ______________________

ALL ABOUT SANDIA HAIRSTREAK BUTTERFLY

LOOK UP THIS BUTTERFLY AND COLOR ACCURATELY AND WRITE DOWN THREE INTERESTING FACTS ABOUT IT.

1. ____

____

2. ____

____

3. ____

____

RESEARCH & DISCOVERIES

USE LIBRARY BOOKS, ENCYCLOPEDIAS OR THE INTERNET TO LEARN MORE.

Color the parts of the world where this animal lives.

DRAW ONE OF MY HOST PLANTS	DRAW MY FOOD	DRAW MY ENEMIES

CREATIVE WRITING

Write a story or poem about the illustration on the right.

MATH TIME

Use this page to practice math!

MOVIE TIME

Watch a video, cartoon or documentary about COMPOSTING.

TITLE:_______________________

____

Draw Your Favorite Scenes:

RATING:

Color the butterfly and add to the background.

COLOR

Use a smooth black marker to complete the drawing

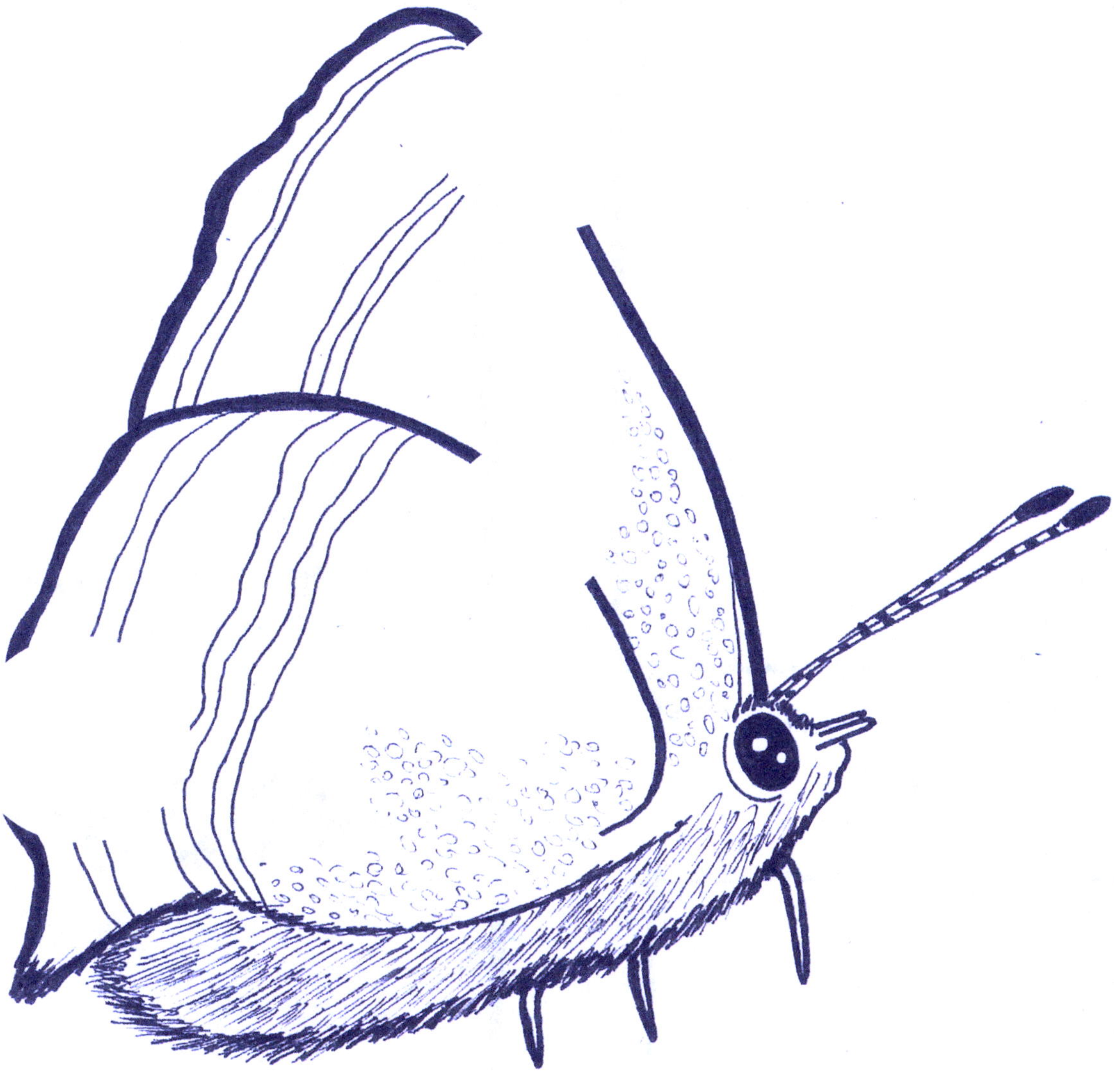

DRAW THE MISSING PARTS

DRAWING TIME

Practice drawing butterflies, bugs, flowers and plants.

Find a way through the maze

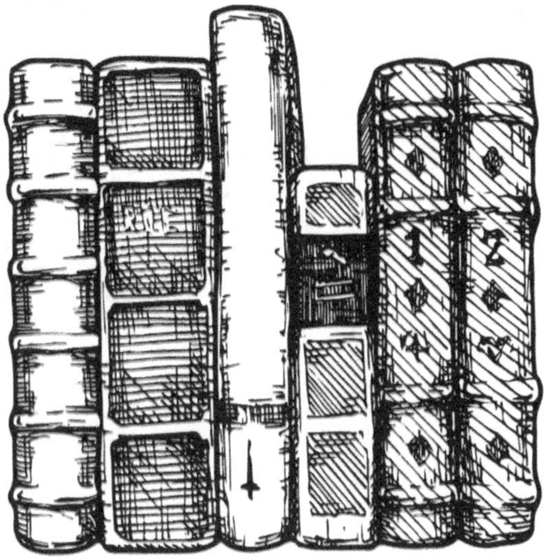

Today I will read for

15 30 45 60

MINUTES

READING TIME

Write and draw about what you are reading.

BACKYARD SCIENCE
NATURE WALK & NATURE STUDY

Draw or write about the things you see outside today.

If possible, find a butterfly or moth to draw!

REDUCE, REUSE, RECYCLE

Draw 5 things that can be put in this bin

Draw garden soil that comes from this bin

REDUCE, REUSE, RECYCLE

Draw a poster that will promote composting.

ALL ABOUT BLACK SWALLOWTAIL BUTTERFLY

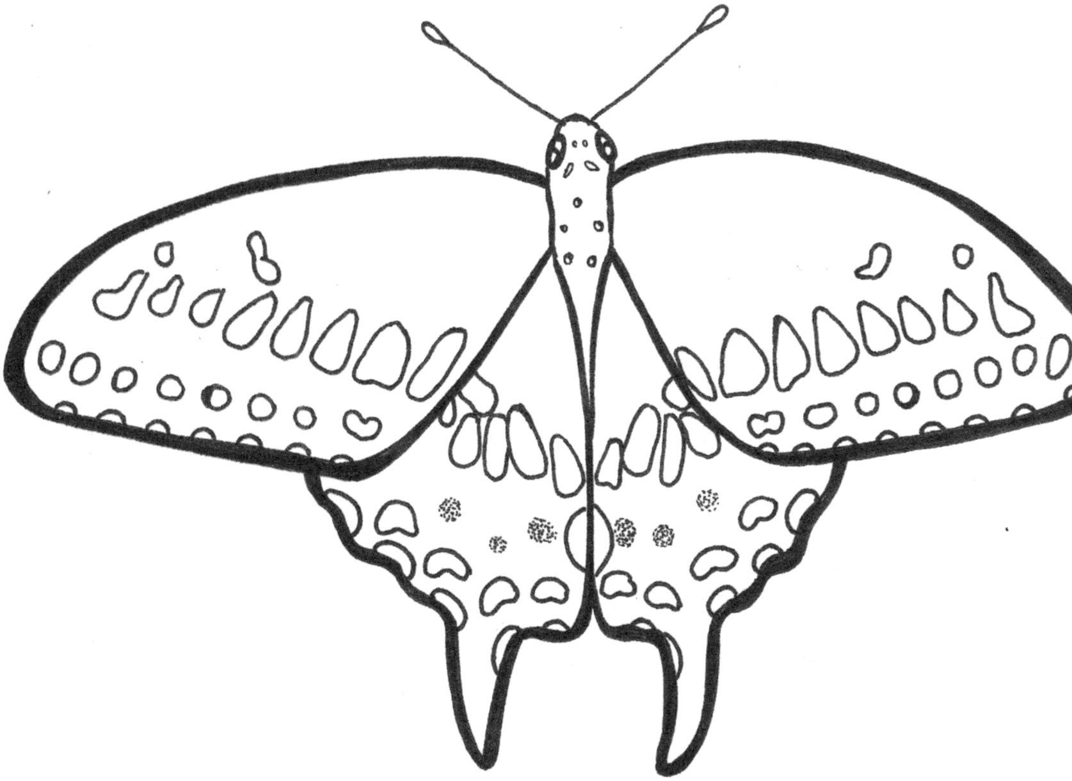

LOOK UP THIS BUTTERFLY AND COLOR ACCURATELY AND WRITE DOWN THREE INTERESTING FACTS ABOUT IT.

1. ____

____

2. ____

____

3. ____

____

RESEARCH & DISCOVERIES

USE LIBRARY BOOKS, ENCYCLOPEDIAS OR THE INTERNET TO LEARN MORE.

Color the parts of the world where this animal lives.

DRAW ONE OF MY HOST PLANTS	DRAW MY FOOD	DRAW MY ENEMIES

CREATIVE WRITING

Write a story or poem about the illustration on the right.

COMPLETE & COLOR

A B C D E F G

H I J K L M

N O P Q R S T

U V W X Y Z

a b c d e f g h

i j k l m n o p

q r s t u v w x

y z 1 2 3 4 5 6 7 8 9 0

CURSIVE WRITING PRACTICE

SPELLING TIME

Pick a Letter ________

Look in your nature and butterfly books for words that start with this letter.

Write ten spelling words.

1.________________________________

2.________________________________

3.________________________________

4.________________________________

5.________________________________

6.________________________________

7.________________________________

8.________________________________

9.________________________________

10.________________________________

LISTENING TIME

IDEAS: MUSIC, PODCASTS, & AUDIO BOOKS

I am listening to: ______________________________

MATH TIME

Use this page to practice math!

MOVIE TIME

Watch a video, cartoon or documentary about RAINWATER HARVESTING.

TITLE:________________________

____

Draw Your Favorite Scenes:

RATING:

CLIP ART

Use stickers, pictures from old magazines, or print out pictures of butterflies and flowers from the internet and paste them on these pages.

COPYWORK

Copy a verse, quote or line of poetry:

TITLE: ______________________

ALL ABOUT BRIMSTONE BUTTERFLY

LOOK UP THIS BUTTERFLY AND COLOR ACCURATELY AND WRITE DOWN THREE INTERESTING FACTS ABOUT IT.

1. ____

____

2. ____

____

3. ____

____

RESEARCH & DISCOVERIES

USE LIBRARY BOOKS, ENCYCLOPEDIAS OR THE INTERNET TO LEARN MORE.

Color the parts of the world where this animal lives.

DRAW ONE OF MY HOST PLANTS	DRAW MY FOOD	DRAW MY ENEMIES

CREATIVE WRITING

Write a story or poem about the illustration on the right.

Color the butterfly and add to the background.

COLOR

Use a smooth black marker to complete the drawing

DRAW THE MISSING PARTS

DRAWING TIME

Practice drawing butterflies, bugs, flowers and plants.

Find a way through the maze

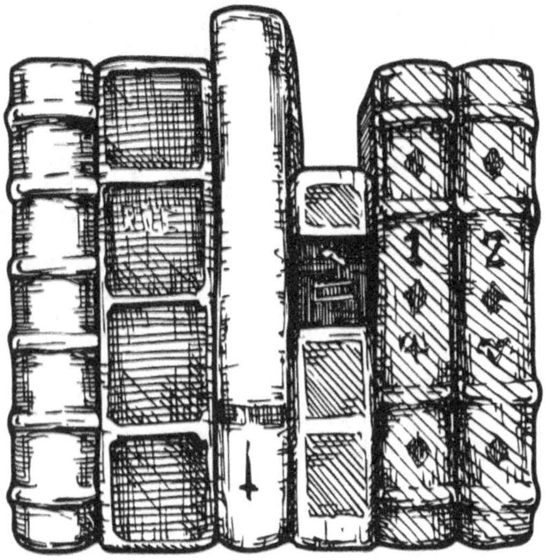

Today I will read for

15 30 45 60

MINUTES

READING

TIME

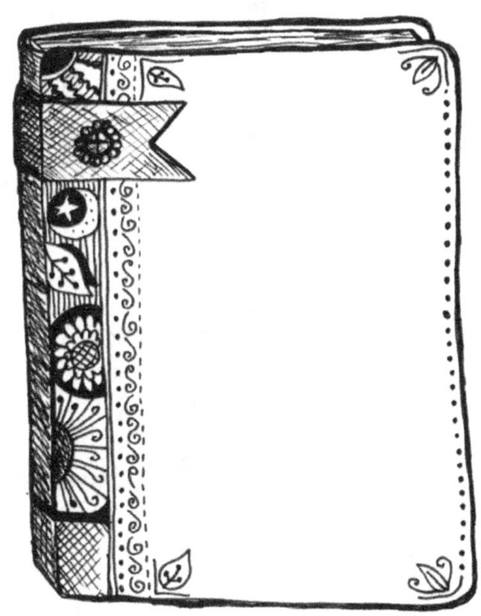

Write and draw about what you are reading.

BACKYARD SCIENCE
NATURE WALK & NATURE STUDY

Draw or write about the things you see outside today.

If possible, find a butterfly or moth to draw!

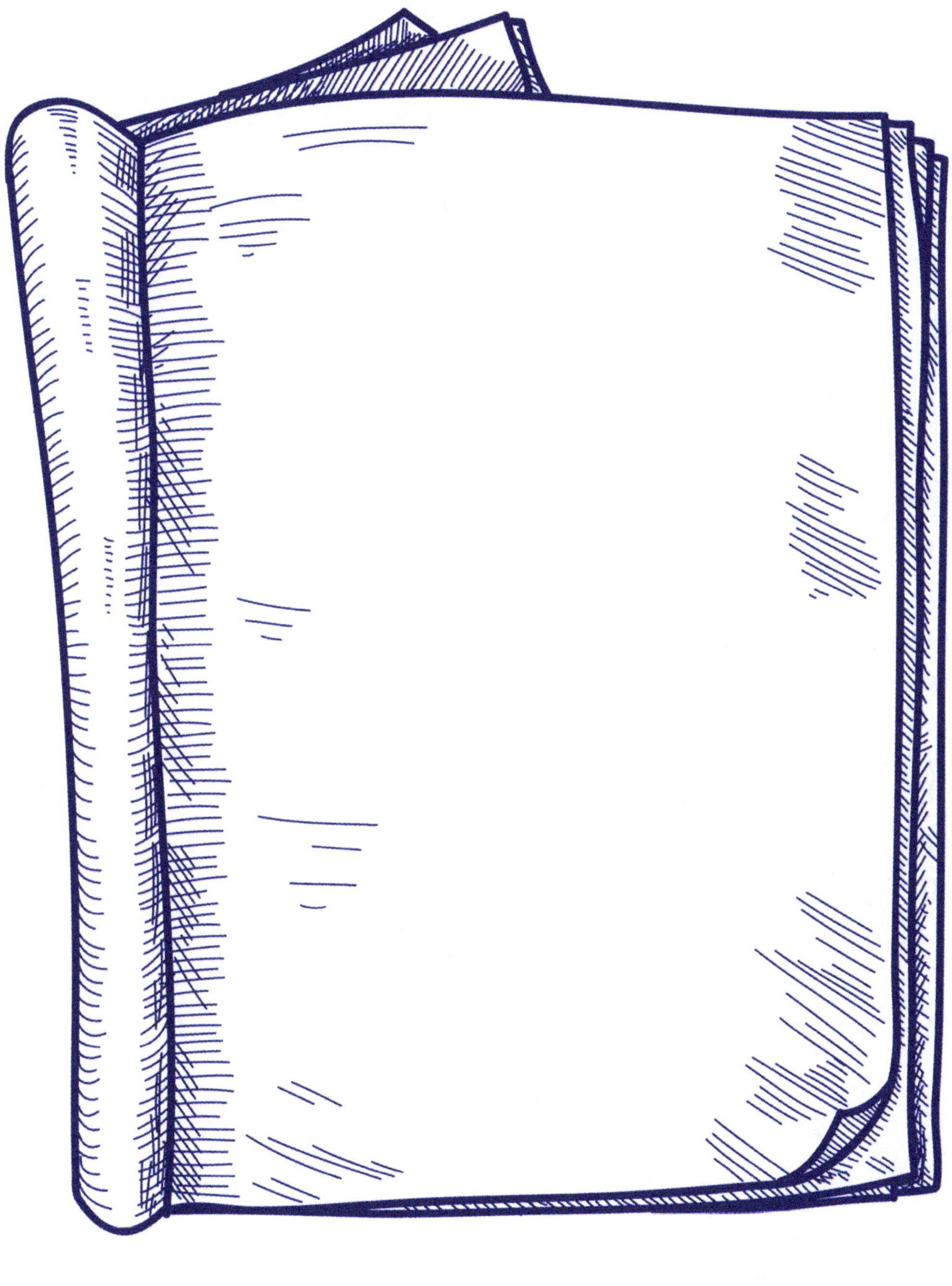

COMPLETE & COLOR

A B C D E F G
H I J K L M
N O P Q R S T
U V W X Y Z

a b c d e f g h
i j k l m n o p
q r s t u v w x
y z 1 2 3 4 5 6 7 8 9 0

CURSIVE WRITING PRACTICE

MATH TIME

Use this page to practice math!

MOVIE TIME

Watch a video, cartoon or documentary about BUTTERFLIES.

TITLE:_________________________

____

RATING:

Draw Your Favorite Scenes:

ALL ABOUT PAINTED LADY BUTTERFLY

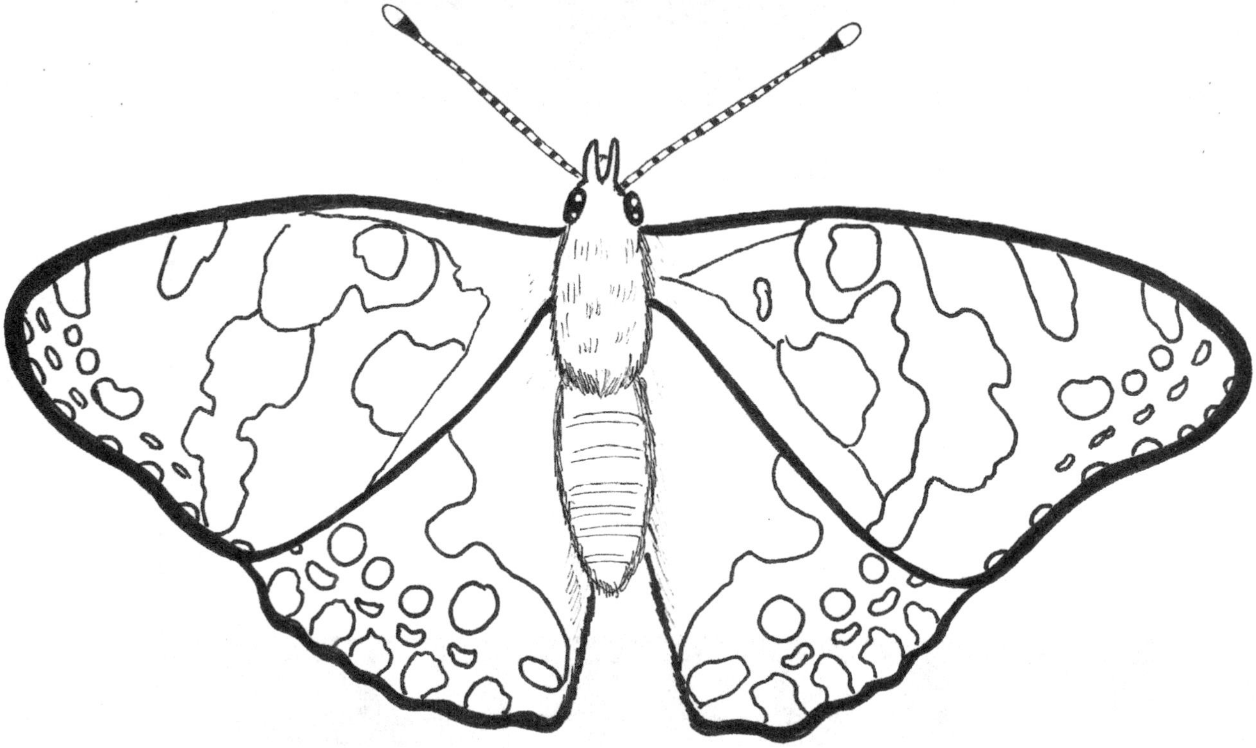

LOOK UP THIS BUTTERFLY AND COLOR ACCURATELY AND WRITE DOWN THREE INTERESTING FACTS ABOUT IT.

1. ____

____

2. ____

____

3. ____

____

RESEARCH & DISCOVERIES

USE LIBRARY BOOKS, ENCYCLOPEDIAS OR THE INTERNET TO LEARN MORE.

Color the parts of the world where this animal lives.

DRAW ONE OF MY HOST PLANTS	DRAW MY FOOD	DRAW MY ENEMIES

CREATIVE WRITING

Write a story or poem about the illustration on the right.

SPELLING TIME

Pick a Letter ________

Look in your nature and butterfly books for words that start with this letter.

Write ten spelling words.

1. ______________________________
2. ______________________________
3. ______________________________
4. ______________________________
5. ______________________________
6. ______________________________
7. ______________________________
8. ______________________________
9. ______________________________
10. ______________________________

LISTENING TIME

IDEAS: MUSIC, PODCASTS, & AUDIO BOOKS

I am listening to: ______________________________

CLIP ART

Use stickers, pictures from old magazines, or print out pictures of butterflies and flowers from the internet and paste them on these pages.

COPYWORK

Copy a verse, quote or line of poetry:

TITLE: ____________________

DRAWING TIME

Practice drawing butterflies, bugs, flowers and plants.

Find a way through the maze

ALL ABOUT
TIGER LONGWING BUTTERFLY

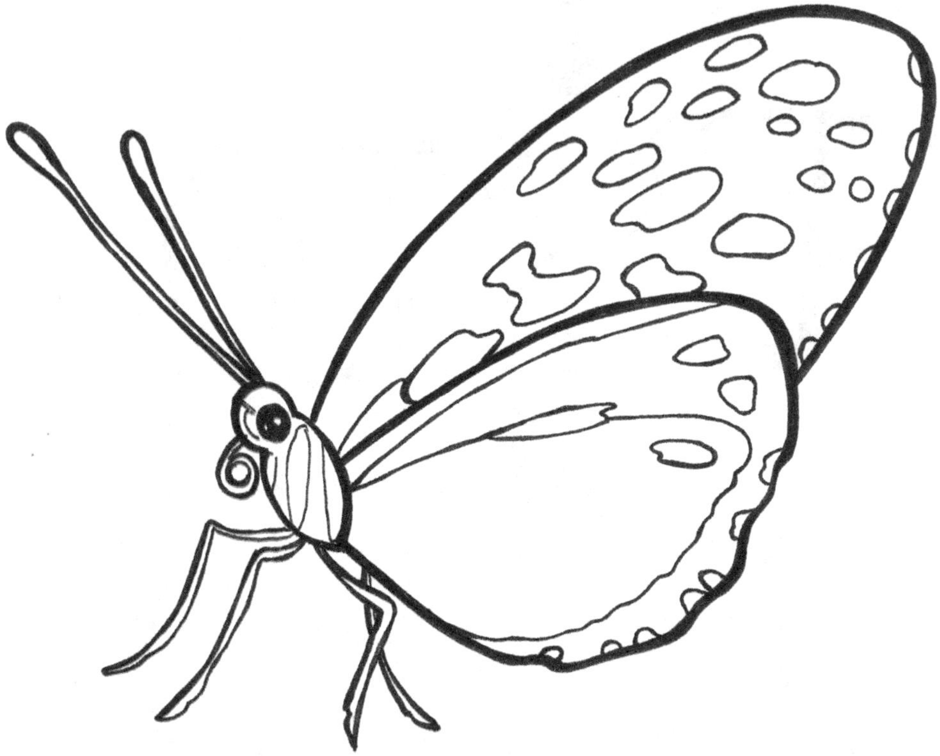

LOOK UP THIS BUTTERFLY AND COLOR ACCURATELY AND WRITE DOWN THREE INTERESTING FACTS ABOUT IT.

1.____

____

2.____

____

3.____

____

RESEARCH & DISCOVERIES

USE LIBRARY BOOKS, ENCYCLOPEDIAS OR THE INTERNET TO LEARN MORE.

Color the parts of the world where this animal lives.

DRAW ONE OF MY HOST PLANTS	DRAW MY FOOD	DRAW MY ENEMIES

CREATIVE WRITING

Write a story or poem about the illustration on the right.

Color the butterfly and add to the background.

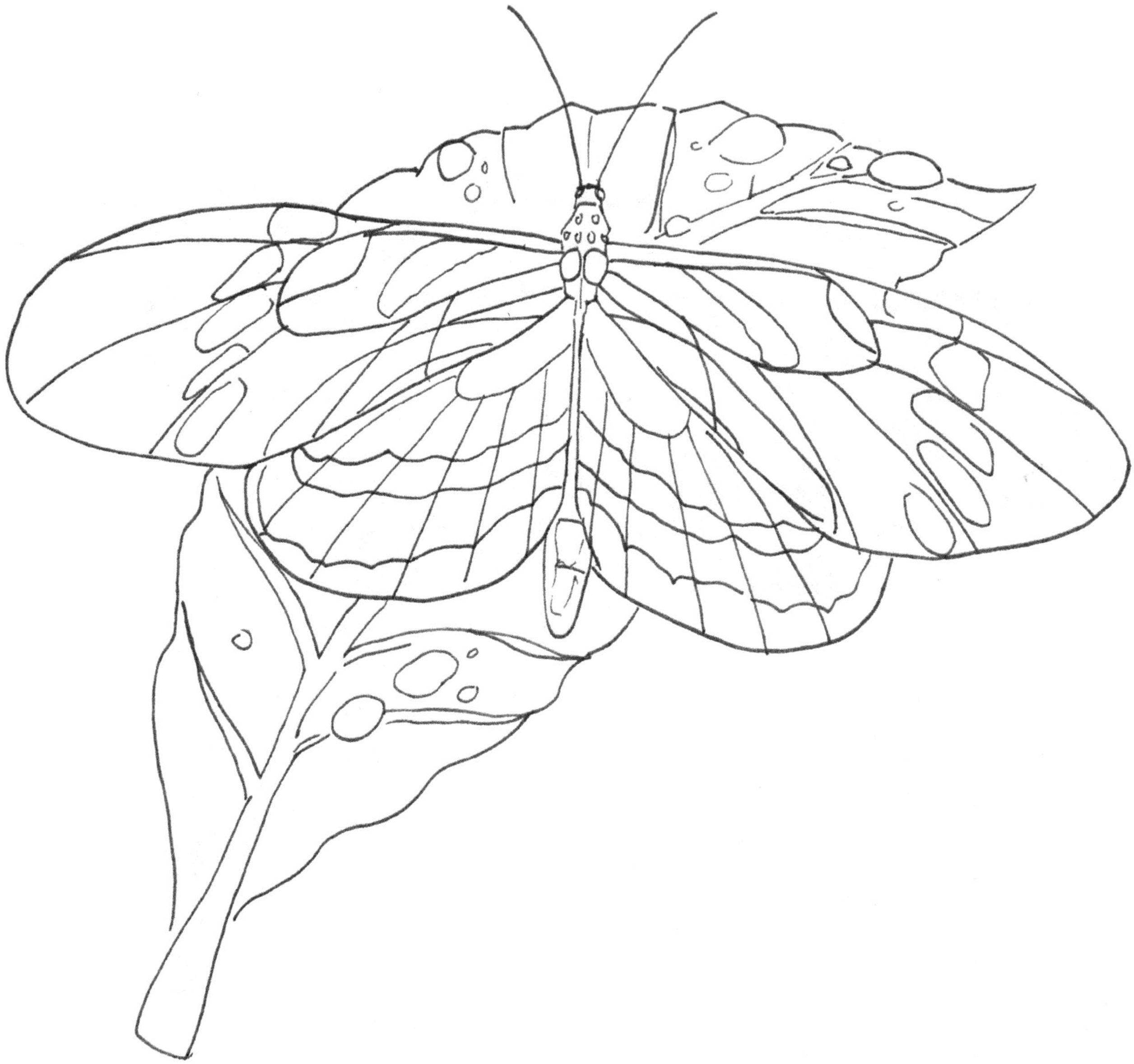

Use a smooth black marker to complete the drawing

DRAW THE MISSING PARTS

REDUCE, REUSE, RECYCLE

Draw 3 things that can be put in this bin

Draw 3 things that can be made of recycled metal

REDUCE, REUSE, RECYCLE

Draw a poster to promote recycling.

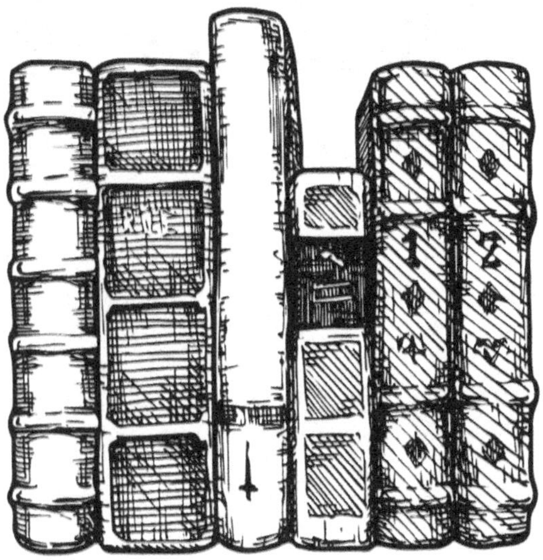

Today I will read for

15 30 45 60

MINUTES

READING TIME

Write and draw about what you are reading.

BACKYARD SCIENCE
NATURE WALK & NATURE STUDY

Draw or write about the things you see outside today.

If possible, find a butterfly or moth to draw!

COMPLETE & COLOR

A B C D E F G
H I J K L M
N O P Q R S T
U V W X Y Z

a b c d e f g h
i j k l m n o p
q r s t u v w x
y z 1 2 3 4 5 6 7 8 9 0

CURSIVE WRITING PRACTICE

MATH TIME

Use this page to practice math!

MOVIE TIME

Watch a video, cartoon or documentary about the importance of REDUCING, REUSING AND RECYCLING.

TITLE:_______________________

____

RATING:

Draw Your Favorite Scenes:

ALL ABOUT QUEEN ALEXANDRA BUTTERFLY

LOOK UP THIS BUTTERFLY AND COLOR ACCURATELY AND WRITE DOWN THREE INTERESTING FACTS ABOUT IT.

1. ______________________________

____

2. ______________________________

____

3. ______________________________

____

RESEARCH & DISCOVERIES

USE LIBRARY BOOKS, ENCYCLOPEDIAS OR THE INTERNET TO LEARN MORE.

Color the parts of the world where this animal lives.

DRAW ONE OF MY HOST PLANTS	DRAW MY FOOD	DRAW MY ENEMIES

CREATIVE WRITING

Write a story or poem about the illustration on the right.

SPELLING TIME

Pick a Letter ________

Look in your nature and butterfly books for words that start with this letter.

Write ten spelling words.

1.________________________________

2.________________________________

3.________________________________

4.________________________________

5.________________________________

6.________________________________

7.________________________________

8.________________________________

9.________________________________

10.________________________________

LISTENING TIME

IDEAS: MUSIC, PODCASTS, & AUDIO BOOKS

I am listening to: ______________________________

CLIP ART

Use stickers, pictures from old magazines, or print out pictures of butterflies and flowers from the internet and paste them on these pages.

COPYWORK

Copy a verse, quote or line of poetry:

TITLE: ____________________

DRAWING TIME

Practice drawing butterflies, bugs, flowers and plants.

Find a way through the maze

ALL ABOUT
FOREST OWL BUTTERFLY

LOOK UP THIS BUTTERFLY AND COLOR ACCURATELY

AND WRITE DOWN THREE INTERESTING FACTS ABOUT IT.

1.____

2.____

3.____

RESEARCH & DISCOVERIES

USE LIBRARY BOOKS, ENCYCLOPEDIAS OR THE INTERNET TO LEARN MORE.

Color the parts of the world where this animal lives.

DRAW ONE OF MY HOST PLANTS	DRAW MY FOOD	DRAW MY ENEMIES

CREATIVE WRITING

Write a story or poem about the illustration on the right.

Color the butterfly and add to the background.

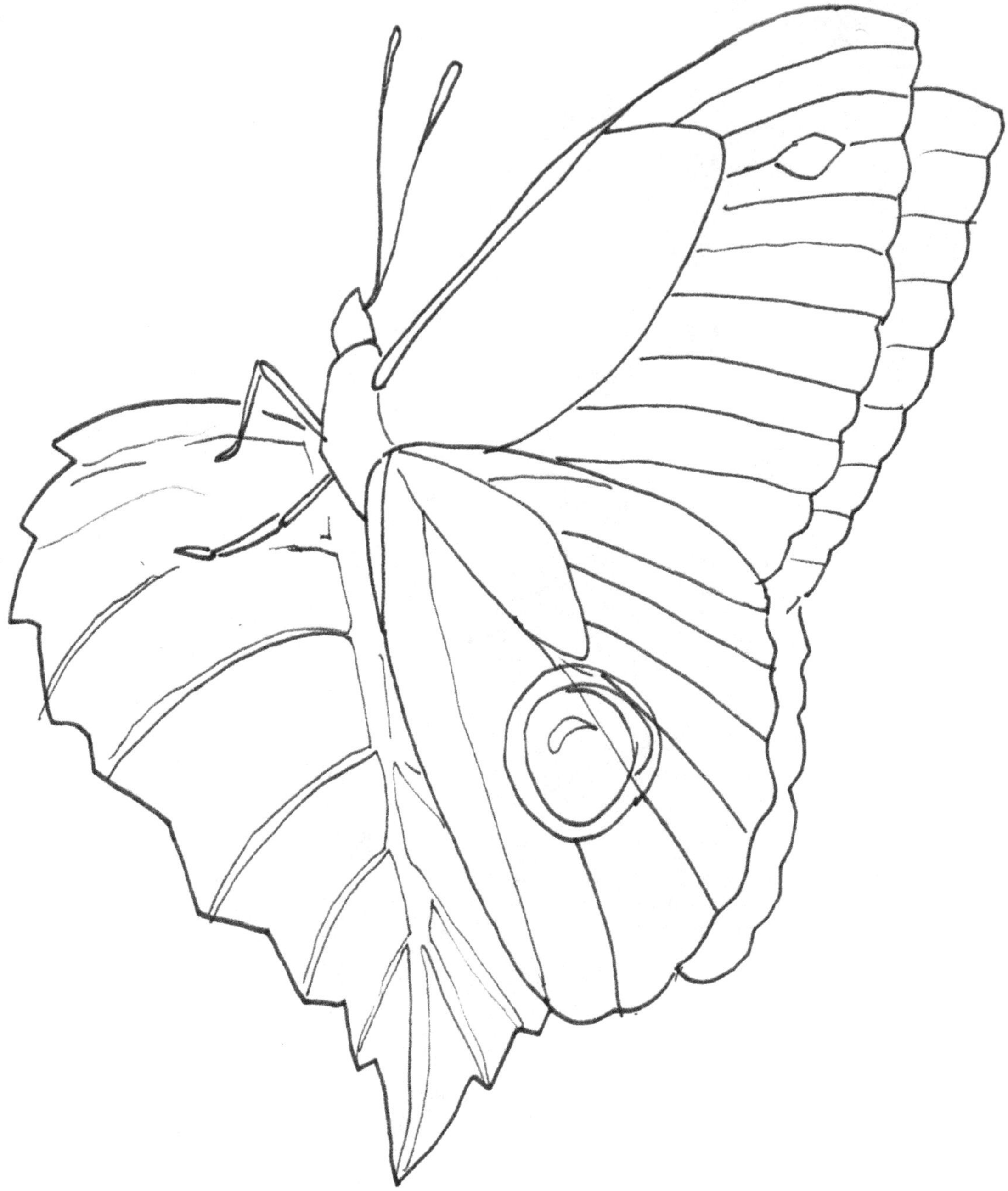

COLOR

Use a smooth black marker to complete the drawing

DRAW THE MISSING PARTS

MATH TIME

Use this page to practice math!

MOVIE TIME

Watch a video, cartoon or documentary about SAVING THE FOREST.

TITLE:______________________

____

Draw Your Favorite Scenes:

RATING:

BACKYARD SCIENCE
NATURE WALK & NATURE STUDY

Draw or write about the things you see outside today.

If possible, find a butterfly or moth to draw!

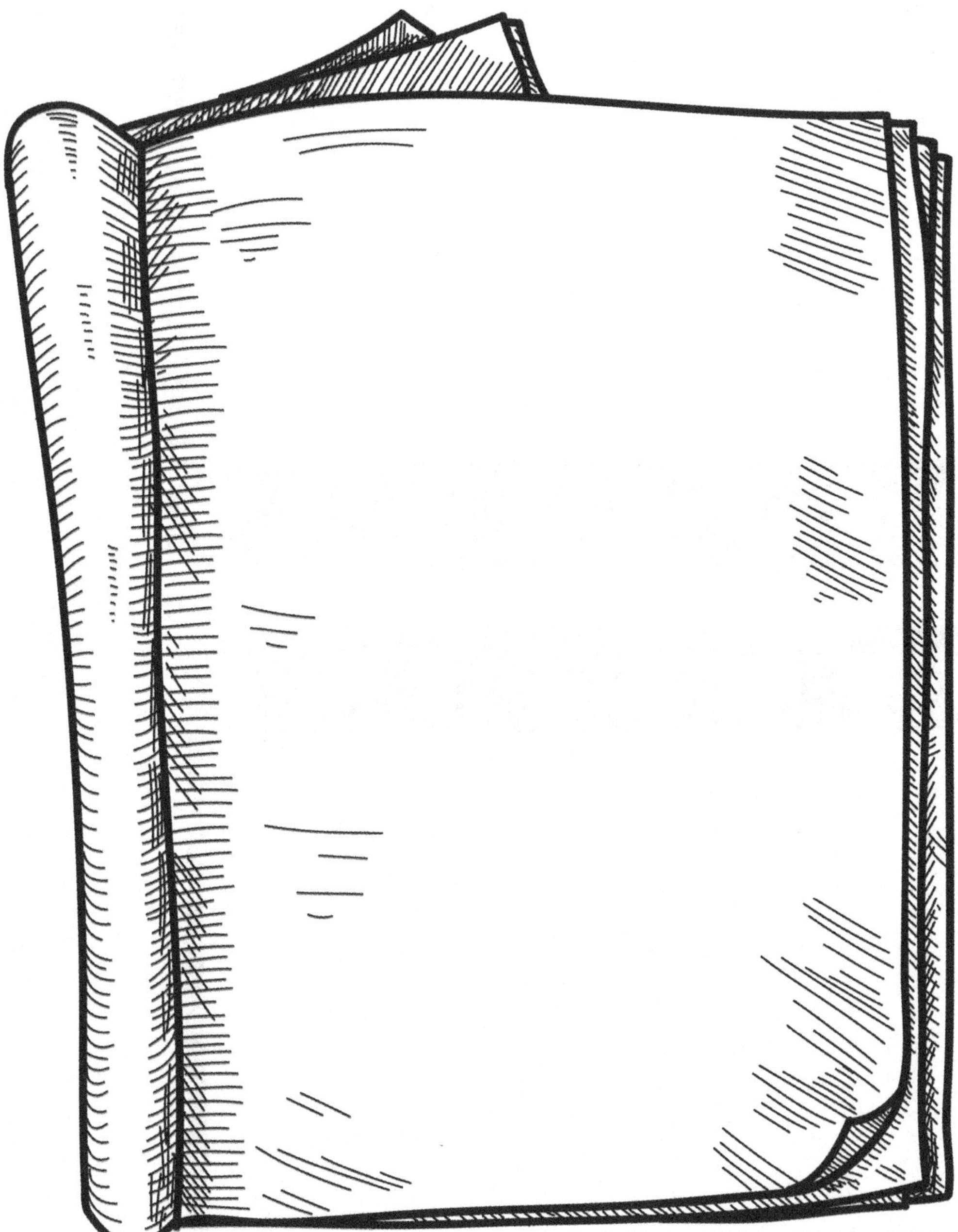

Today I will read for

15 30 45 60

MINUTES

READING TIME

Write and draw about what you are reading.

COMPLETE & COLOR

A B C D E F G
H I J K L M
N O P Q R S T
U V W X Y Z

a b c d e f g h
i j k l m n o p
q r s t u v w x
y z 1 2 3 4 5 6 7 8 9 0

CURSIVE WRITING PRACTICE

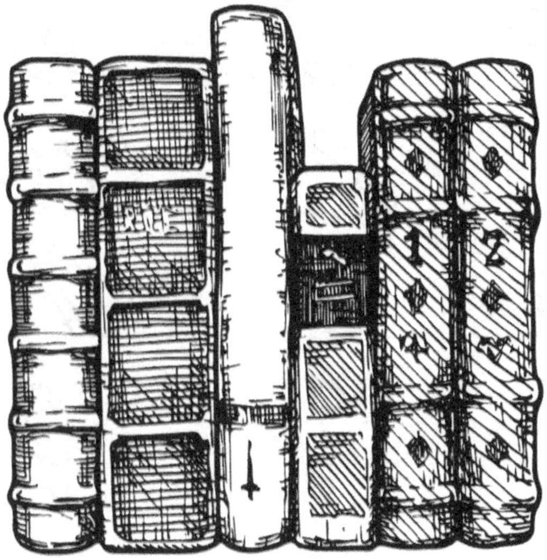

Today I will read for

15 30 45 60

MINUTES

READING

TIME

Write and draw about what you are reading.

LIBRARY BOOKS

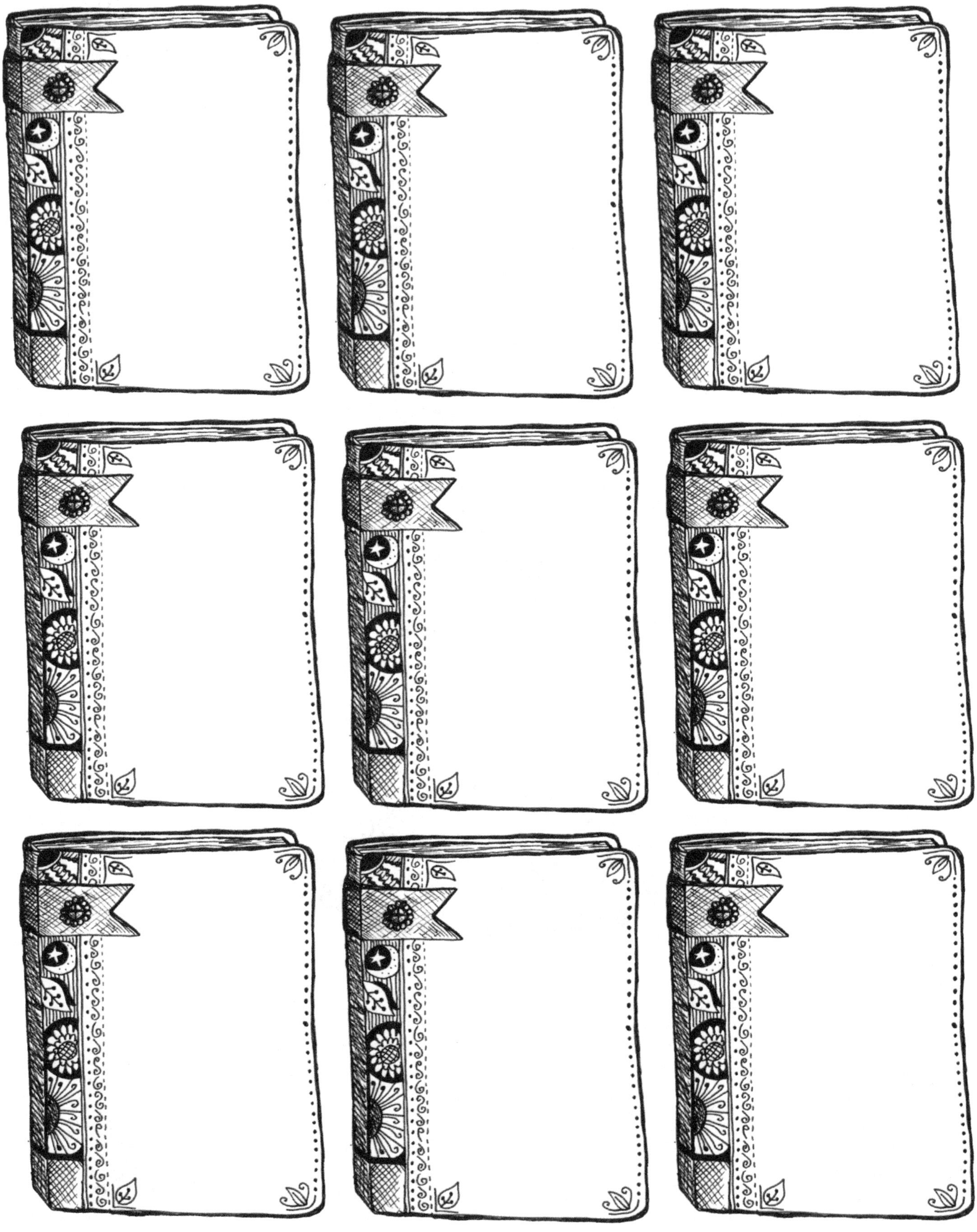

MY LEARNING LIST

Additional Books & Documentaries

TITLE: DATE:

LIBRARY BOOKS

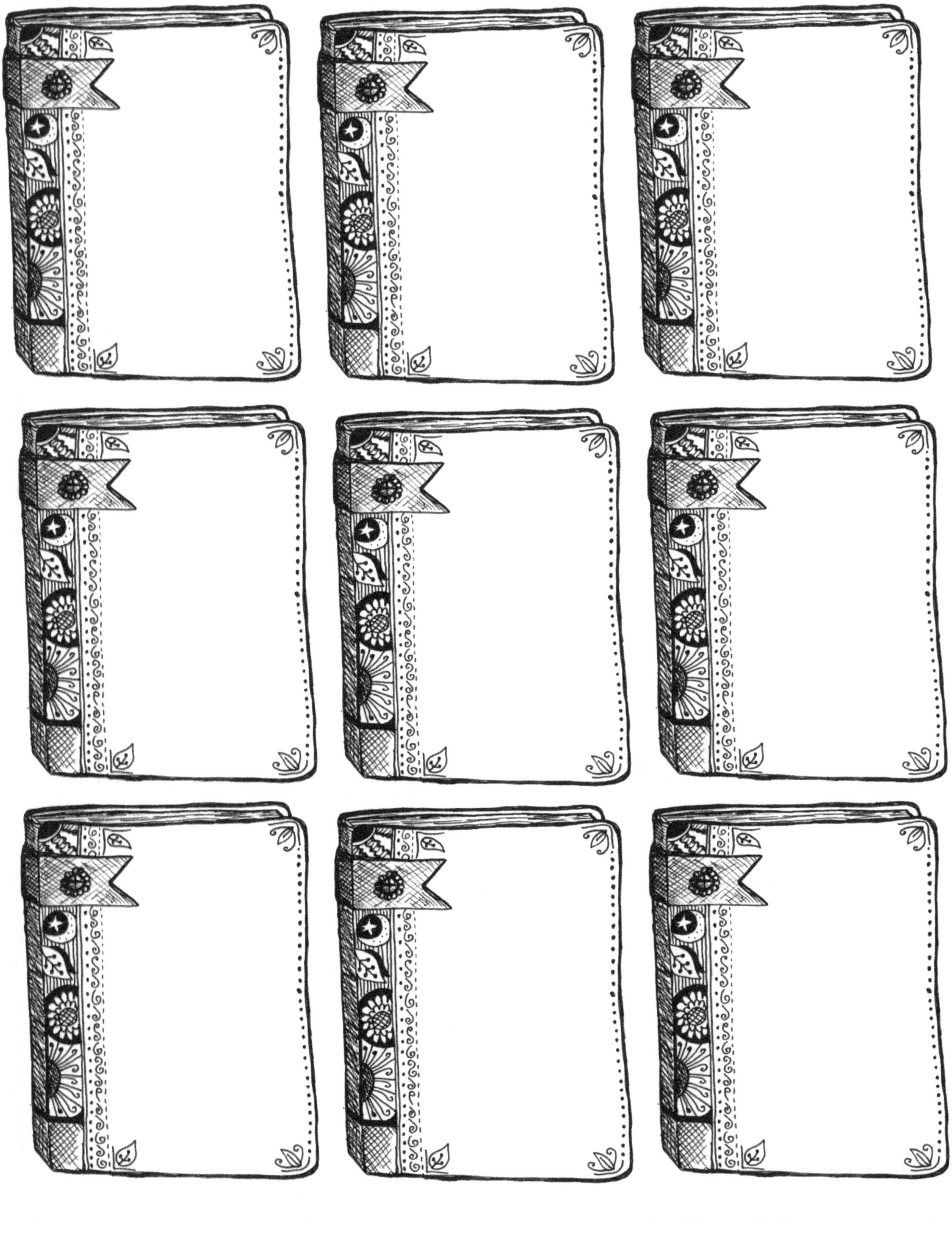

ALL ABOUT FUN-SCHOOLING

A Learning Lifestyle by The Thinking Tree

We exist to make learning fun for everyone by designing tools and experiences that spark curiosity and set people free to learn through their passions—because the world needs bold, curious thinkers, not cookie-cutter students.

For parents and educators who feel traditional methods aren't working, Fun-Schooling offers a refreshing alternative: interest-based, neurodivergent-friendly journals and workbooks that teach children how to learn—joyfully, creatively, and confidently.

OUR STORY

Fun-Schooling began when artist and mother of fifteen, Sarah Janisse Brown, needed a better way to teach her creative, outside-the-box learners. Some children loved art. Some loved animals. Some had dyslexia or attention challenges. Each child learned differently—and standardized lessons simply couldn't meet their needs.

So Sarah began designing personalized learning journals that followed the child instead of forcing a rigid curriculum. These journals encouraged curiosity, creativity, and independence by starting with a child's interests, strengths, and questions.

Families around the world soon discovered the same truth: **children learn best when their passions lead the way.**

Today, The Thinking Tree publishes journals and workbooks for every age and learning style. Our heartbeat remains the same: **Honor the child. Follow their spark. Let learning become an adventure.**

WHY CHOOSE FUN-SCHOOLING?

Every child has a different style of learning. We recognize five distinct types of learners —Detectives, Creators, Followers, Friends, and Explorers. These learning styles are simply guides—not labels—and many children move between them as they grow.

Fun-Schooling works with these natural styles — rather than against them — by nurturing curiosity, independence, confidence, and a lifelong love of learning. Children are encouraged to:

- choose the books they want to learn from
- explore interests deeply and at their own pace
- express learning through drawing, writing, doodles, lists, maps, recipes, experiments, and more

This approach teaches students **how to learn**, not just what to memorize.

HOW FUN-SCHOOLING BOOKS WORK

Fun-Schooling books mirror how learning happens in real life.

Children naturally build their education using Fun-Schooling books alongside:

- library books
- books they already own
- documentaries & educational videos
- field trips & nature experiences
- projects and hobbies
- simple internet research (with guided questions and parent supervision)

We intentionally steer children away from passive screen dependence while embracing technology in a thoughtful, balanced way.

There is no one-size-fits-all approach. Each journal blends multiple subjects — often including science, social studies, and language arts — so learning feels connected and meaningful.

Every Fun-Schooling book becomes a customized, child-led learning experience. Parents love that a single book can be used across a wide age range. Siblings can work together, making learning collaborative, flexible, and simple.

With over 300 different books, Fun-Schooling keeps each child's interests and giftings at the center of their education.

FUN-SCHOOLING BOOK TYPES

BOOK TYPE	STRUCTURE	OUTSIDE RESOURCES UTILIZED?	EXAMPLES
Core Journal	Multisubject, open-ended framework	Yes — 5–9 books chosen by the student or educator	Library books, home books, documentaries, nature walks, simple research
Themed Journal	Open-ended within a specific theme	Yes — varies by topic	Topic-specific books, documentaries, museum visits, guided online research
Handbook	Guided research with structured prompts	Sometimes	Reference books, reading books, real-world observations
Workbook	Self-contained guided practice	Rarely	Occasionally 1–2 specific books, optional reference materials
Activity Book	Hands-on puzzles, mazes, projects, challenges	Not usually	Simple household supplies
Coloring Book	Creative, calming, open-ended	No outside resources	Only art supplies

WHAT MAKES FUN-SCHOOLING DIFFERENT?

Fun-Schooling is not a curriculum you must follow.

It's a framework that adapts to each family or classroom.

It works especially well for:

- creative and out-of-the-box thinkers
- struggling or discouraged learners
- neurodivergent students
- families or educators seeking freedom, flexibility, and joy

We don't teach children to memorize and repeat. We teach them to think, explore, and engage with the world as their classroom.

When learning is driven by curiosity, passion, and resourcefulness—children thrive.

HOW TO FUN-SCHOOL

Most Fun-Schooling families begin by choosing a bundle of Thinking Tree books that match their child's age and interests.

1. Read the introductions in each journal for simple guidance and suggested materials.
2. Set up two baskets: one for library books and one for journals and supplies.
3. Create a cozy learning space—then take your baskets outside or on the go.
4. Learn like real life: read, watch, explore, build, draw, write, and research while following the prompts inside each Fun-Schooling book.

Your role isn't to have all the answers—it's to provide resources, ask good questions, and let curiosity lead. For homeschooling parents, we encourage Mom-Schooling too. Choose a journal of your own, gather books on topics you love, and model focused, joyful learning alongside your children. Often, children learn more from what you do than from what you teach.

That's it: follow curiosity, gather resources, lead by example, and capture learning—joyfully.

WELCOME TO THE MOVEMENT

Fun-Schooling is more than a method. It's a mindset, a community, and a learning lifestyle.

Thousands of families worldwide are choosing joy over pressure, creativity over conformity, and curiosity over standardization. You don't need to recreate school at home—you just need to create space for curiosity to grow.

Illustrations by Sarah Janisse Brown, Notika Pashchenko & Mid Journey

Made in the USA
Coppell, TX
07 September 2026

89263309R00125